T0076647

Biological Nanostructures and Applications of Nanostructures in Biology

Electrical, Mechanical, and Optical Properties

BIOELECTRIC ENGINEERING

Series Editor: Bin He
University of Minnesota
Minneapolis, Minnesota

MODELING AND IMAGING OF BIOELECTRICAL ACTIVITY
Principles and Applications
Edited by Bin He

BIOLOGICAL NANOSTRUCTURES AND APPLICATIONS OF NANOSTRUCTURES IN BIOLOGY
Electrical, Mechanical, and Optical Properties
Edited by Michael A. Stroscio and Mitra Dutta

NEURAL ENGINEERING
Edited by Bin He

Biological Nanostructures and Applications of Nanostructures in Biology

Electrical, Mechanical, and Optical Properties

Edited by

Michael A. Stroscio and Mitra Dutta

University of Illinois at Chicago
Chicago, Illinois

Kluwer Academic / Plenum Publishers
New York, Boston, Dordrecht, London, Moscow

ISBN 0-306-48627-X

233 Spring Street, New York, New York 10013

http://www.wkap.nl/

10 9 8 7 6 5 4 3 2 1

A C.I.P. record for this book is available from the Library of Congress

PREFACE

Nanostructures are playing a fundamental role in the advancement of biology as a result of the continuing dramatic progress in understanding the electronic, optical, and mechanical properties of an ever increasing variety of nanostructures. This book on "Biological Nanostructures and Applications of Nanostructures in Biology: Electrical, Mechanical, and Optical Properties" highlights recent past advances at the interface between the science and technology of nanostructures and the science of biology. Moreover, this book supplements these past groundbreaking discoveries with discussions of promising new avenues of research that reveal the enormous potential of emerging approaches in nanobiotechnology.

A dominant trend of the last two decades has been down scaling of electronic, optoelectronic, and mechanical devices and structures from micron scale to those with features into the nano-dimensional regime. In most cases, this downscaling has not been possible without taking into account the dramatic differences between micron-scale phenomena that may be described frequently in terms of classical theories and nanoscale phenomena that generally require the consideration of quantum nature of matter. Moreover, many new technologies – such as those of heteroepitaxy, synthesis of carbon nanotubes, and synthesis of nanostructures through colloidal chemistry – have emerged and been developed. In recent years, these developments have been recognized as offering powerful tools in the quest to advance the basic understanding of biological systems as well as in biomedical applications of nanotechnology. As an example of a widespread application of nanotechnology, the diversity of heterojunctions as well as dramatic advances in semiconductor growth and processing technologies are opening the way to new heterojunction-device technologies and leading to many new avenues for realizing novel families of quantum-based electronic and optoelectronic devices and systems. Even more important from the perspective of biological applications, the already-large number of applications of advanced semiconductor heterostructures is increasing rapidly and is becoming more diversified as illustrated by the wide range of uses of layered quantum dots in biological applications. As highlighted in this book, these applications include using quantum dots for biological tags as well as for active optical and electrical interface with biological systems such as neurons.

Indeed, advanced semiconductor heterostructures can be expected contribute to revolutionary advances in medical applications as discussed in this book.

This book highlights current advances and trends in the use of semiconductor nanocrystals in biological applications as well as key developments in the synthesis and physical properties of semiconductor nanocrystals used in biological environments. The potential applications of these semiconductor nanocrystals in nanobiotechnology have been demonstarted recently by a broad variety of applications in the study of subcellular processes of fundamental importance in biology. Examples include the use of colloidal nanocrystals to study neural processes on the nanoscale through the imaging of the diffusion of glycine receptors as well as multi-color labeling of subcellular components. As discussed in this book, semiconductor nanocrystals have narrow, tunable and symmetric emission spectra, and they have much greater temporal stability and resistance to photobleaching than fluorescent dyes.

This book also highlights recent findings on how the electrical, optical, and mechanical properties of nanostructures are altered as a result of being in close proximity with biological structures and media. For example, this book discusses how electrolytic environments, that are pervasive in biological systems, must be considered in understanding the electrical and optical properties of charged and polar semiconductor quantum dots in electrolytic environments. As a second example, this book discusses how dielectric environments, that are pervasive in biological systems, must be considered in understanding the electrical, mechanical, and optical properties of charged and polar semiconductor quantum dots in dielectric media. Moreover, this book highlights the nanomechanical properties of biomolecules in biological environments. Accordingly, this book provides an introduction to a range of topics dealing with the control and tailoring of the properties of both manmade and naturally occurring nanostructures in biological environments. This field is in its infancy but, as indicated by the contraubtions in this book, it is critical to exploiting fully the dramatic advances of nanotechnology in biological and biomedical applications.

This book also describes a number of other promising – and potentially revolutionary – tools and applications of nanobiotechnology. These include: the potential applications of nanoscale carbon nanotubes in bioengineering; the use of atomic force microscopy to probe the nanophysical properties of living cells; and bioinspired approaches to building nanostructures. As described in this book, manmade carbon nanotubes have a number of remarkable electrical, chemical, and mechanical properties making them intriguing candidates for integration with biological structures on the nanoscale. Moreover, this book provides an introduction to potentially revolutionary applications of a key set of biochemical interactions for the assembly and building of nanostructures.

The guest editors wish to acknowledge professional colleagues, friends and family members whose contributions and sacrifices made it possible to complete this work. First of all, the authors are grateful Aaron Johnson of Kluwer Publishing Company for taking an active interest in making this volume useful to the expected readership. The guest editors extend sincere thanks go to Dean Larry Kennedy, College of Engineering, University of Illinois at Chicago (UIC) for his active encouragement and for his longstanding efforts to promote excellence in research at UIC. Special thanks go to Dr. Rajinder Khosla, Dr. James W. Mink and Usha

Varshney of the National Science Foundation, Dr. Dwight Woolard of the US Army Research Office, Dr. John Carrano of the Defense Advanced Research Projects Agency, and Dr. Todd Steiner and Dr. Daniel Johnstone of the Air Force Office of Scientific Research for their encouragement and interests. MD acknowledges the discussions, interactions and the work of many colleagues and friends which have had an impact on the work in this book. Drs. Doran Smith, K. K. Choi and Paul Shen of the Army Research Laboratory, and Professor Athos Petrou of State University of New York, Buffalo. MD would also like to thank Dhiren Dutta without whose encouragement she would never have embarked on a career in science and to Michael and Gautam Stroscio who everyday add meaning to everything. MAS acknowledges the essential roles that several professional colleagues and friends played in the events leading to his contributions to this book; these people include: Professor Bin He of the University of Minnesota, Professor Richard L. Magin, Head of the BioEngineering Department at the University of Illinois at Chicago (UIC), Professors Jeremy Mao and Anjum Ansari of UIC, Professor Duan P. Chen of Rush University, Professors Robert Trew, Gerald J. Iafrate, M. A. Littlejohn K. W. Kim, R. and M. Kolbas as well as Dr. Sergiy Komirenko of the North Carolina State University, Professor Vladimir Mitin of the State University of New York at Buffalo, University, Professors G. Belenky and S. Luryi and Dr. M. Kisin of the State University of New York at Stony Brook, Professors George I. Haddad, Pallab K. Bhattacharya, and Jasprit Singh and Dr. J.-P. Sun of the University of Michigan, Professors Karl Hess and J.-P. Leburton University of Illinois at Urbana-Champaign, Professor L. F. Register of the University of Texas at Austin, Professors H. Craig Casey and Steven Teitsworth of Duke University, and Professor Viatcheslav A. Kochelap of the National Academy of Sciences of the Ukraine. MAS also thanks family members who have been supportive during the period when this book was being edited; these include: Anthony and Norma Stroscio, Mitra Dutta, and Elizabeth, Gautam, and Marshall Stroscio.

CONTENTS

INTEGRATING AND TAGGING BIOLOGICAL STRUCTURES WITH NANOSCALE SEMICONDUCTOR QUANTUM DOT STRUCTURES

Michael A. Stroscio, Mitra Dutta, Kavita Narwani, Peng Shi, Dinakar Ramadurai, Babak Kohanpour, and Salvador Rufo*

1. INTRODUCTION

This account highlights current trends in the use of semiconductor nanocrystals in biological applications as well as key developments in the synthesis and physical properties of semiconductor nanocrystals used in biological environments. The potential applications of semiconductor nanocrystals in nanobiotechnology have been highlighted recently by a broad variety of applications[1] in the study of subcellular processes of fundamental importance in biology. In recent years, semiconductor nanocrystals have been synthesized and evaluated for their potential as fluorescent biological labels. As is now well recognized,[2,3] semiconductor nanocrystals have narrow, tunable and symmetric emission spectra, and they have much greater temporal stability and resistance to photobleaching than fluorescent dyes. Furthermore, fluorescent semiconductor nanocrystals may be bound to biomolecules to facilitate selective binding of these nanoscale[1-8] fluorescent structures to specific subcellular structures. Semiconductor

* Michael A. Stroscio, Depts. of Bioengineering, Electrical and Computer Engineering, and Physics, Univ of IL at Chicago, Chicago, IL 60607. Mitra Dutta, Depts. of Electrical and Computer Engineering, and Physics, Univ of IL at Chicago, Chicago, IL 60607. Babak Kohanpour and Salvador Rufo, Dept. of Electrical and Computer Engineering, and Physics, Univ of IL at Chicago, Chicago, IL 60607. Kavita Narwani, Dinakar Ramadurai, and Peng Shi, Dept. of Bioengineering, Univ of IL at Chicago, Chicago, IL 60607.

nanocrystals find many applications in biology because their emission spectra may be tuned merely by changing their diameters.[9] Moreover, the simultaneous use of semiconductor nanocrystals with a variety of emission spectra opens the way to using multiplexed optical coding in studying complex biological systems. In related efforts, the programmed assembly of DNA functionalized semiconductor nanocrystals[10] has been accomplished and techniques for functionalizing the surfaces of gold nanoparticles[11,12] have proven to be of great utility. The use of peptide-functionalized semiconductors for the tagging of cells has been demonstrated widely as will be discussed in this article. Illustrative examples include the tagging of cancer cells,[13] and nerve cells.[14,15] In these works, known methods for synthesizing colloidal suspensions of semiconductor nanocrystals[16] were applied in conjunction with cysteine-based binding of peptides to nanocrystals. The short-peptide sequences used in these works ensure that the quantum dots bind in close proximity to the cell and recognition-molecule-directed interfacing between semiconductor quantum dots to cellular integrins is accomplished by using RGD, RGDS, IKVAV, and other peptide sequences. In addition to emphasizing biological applications of nanocrystals, this article highlights the electrical and optical properties of these semiconductor nanocrystals. In recent years, a number of reviews on the synthesis,[17] size- and shape-dependent properties,[18] characterization,[19] and electron-phonon interactions,[20] of these nanocrystals have been published; indeed, there are also several books available that emphasize the theory and experimental characterization of semiconductor nanocrystals.[9,21-23] The basic physical properties of semiconductor nanocrystals have been treated in detail; treatments include techniques for Raman scattering from arrays of semiconductor nanocrystals,[24] optical lattice vibrations of polar semiconductor quantum dots,[25] the effective-phonon approximation of polarons in ternary mixed crystals,[26] and models of optical linewidths of individual quantum dots.[27] Moreover, the basic electrical, optical, and mechanical properties of GaAs-based semiconductor nanocrystals have been studied extensively.[28-38] The many applications of semiconductor nanocrystals include single-photon detection in the far-infrared portion of the electromagnetic spectrum.[39] Moreover, the electrical, optical, and mechanical properties of InAs-based semiconductor nanocrystals have been studied extensively.[40-49] Applications include: (a) mid-infrared second-harmonic generation in p-type InAs/GaAs self-assembled quantum dots,[50] (b) self-assembled InAs/GaAs quantum dot intersubband detectors,[51] (c) InAs-based infrared photodetectors,[52,53] (d) mid-infrared absorption and photocurrent spectroscopy of InAs/GaAs self-assembled quantum dots,[54] (e) InAs quantum dot field effect transistors,[55] and (f) InGaAs/GaAs quantum dot lasers.[56] Many other semiconductor nanocrystal systems have been studied; these include: GaSb/GaAs,[57] GaN,[58,59] PbS,[60-64] PbSe,[65,66] CdTe,[67-69] InP,[70-73] and CdS.[74] Imada et al.[75] have given a detailed photoreflectance study of bulk CdS that provides the exciton binding energy for bulk CdS; in particular, this paper provides detailed information on the temperature-dependent bandgap as well as the excitonic properties of CdS. The CdS exciton binding energy is reported to be about 30 meV. The study of CdSe-based nanocrystals has been extensive;[76-84] these studies include characterization and synthesis efforts as well as investigations of the optical and electrical properties.[85-106] Specialized studies focus on CdSe-based single-electron transistors[107] as well as on infrared[108-109] and magnetic properties.[110] As is well known, an understanding of the properties of semiconductor quantum dots has been critical to the realization of quantum-dot lasers.[111-113] The insights on the electronic and optical properties of nanocrystals gained in these studies provide a foundation for understanding related efforts related to the colloidal nanocrystals of primary interest in biological applications.

2. FABRICATING QUANTUM DOT SYSTEMS AND THEIR APPLICATIONS AS BIOTAGS

As discussed previously, semiconductor nanocrystals have been prepared and studied experimentally to assess their utility as fluorescent biological labels having narrow, tunable and symmetric emission spectra with properties that are potentially superior to those of conventional dyes. Many of these nanocrystals are fabricated from II-VI semiconductor materials. Table 1 summarizes the energy bandgaps for selected III-V, II-

Compound Semiconductor	Bandgap, (eV) (direct bandgaps unless otherwise specified)	Spontaneous Polarization, (C/m^2)
AlN		- 0.081[114]
CdS hexagonal CdS cubic	2.4 E_g (A) 2.5 E_g (B) 2.55 E_g (C) 2.5	0.002[115]
CdSe hexagonal CdSe cubic	1.75 E_g (A) 1.771 E_g (B) 2.17 E_g (C) 1.9	0.006[89]
CdTe	1.49	
PbS	0.41	
PbSe	0.27	
ZnS	3.68	
GaN	3.36	- 0.029[114]
ZnO	3.35	- 0.007[115]

Table 1. Energy bandgaps and spontaneous polarizations of selected compound semiconductors.

VI, and IV-VI semiconductors as well as the spontaneous polarizations for selected II-VI semiconductors. Among these, CdSe and CdS nanocrystals have been prepared in colloidal suspensions as discussed in this article. GaN nanocrystals with diameters in the range of 1.6 nm to 4.5 nm have been synthesized using laser-ablation of Ga into a nitrogen atmophere.[116] As will be discussed later, the spontaneous polarization of these würtzite structure II-VI materials is an intrinsic property of these polar materials. This spontaneous polarization produces an internal electric field in the nanocrystal. As in all semiconductors, such a field results in a slope in the conduction bandedge $E_c(x)$, given by $dE_c(x)/dx = q\,E$ where E is the electric field produced by the spontaneous polarization.

Recently, great strides have been made in the synthesis and functionalization of nanocrystals. As discussed by Wehrenberg et al.[65] lead sellenide (PbSe) colloidal nanocrystals are well-suited as nanoscale biotags in the infrared region of the spectrum from about 0.5 to 1.0 eV. The use of PbSe quantum dots as fluorescent biotags is especially promising since infrared organic dyes are very poor fluorophors. Moreover, biological tissues are relatively transparent in the near IR-spectral range. As indicated by Wehrenberg et al., nanocrystals of highly monodisperse PbS and PbSe colloidal nanocrystals --- that is, with small variation in the range of quantum-dot diameters --- have been prepared via the techniques of colloidal chemistry. In particular, Wehrenberg et al. have prepared PbSe quantum dots with a capping of oleic acid at moderate temperatures in organic solvents, leading to monodisperse samples. Two solutions were employed to achieve these results: (a) a solution made of phenyl ester (2 mL), oleic acid (1.5 mL) and trioctylphosphine (8mL), and dissolved in lead (II) acetate trihydrate (.65g); and (b) another solution contains 10mL of phenyl ester. Both solutions were heated under vacuum for an hour at 85° C. Solution (a) was then cooled under an atmosphere of inert argon to a temperature of 45°C. Solution (b) containing phenyl ester was heated to a temperature between 180 and 210° C under an inert atmosphere of argon. 1.7 mL 1M Trioctylphosphine (TOPSe) was then added to solution (a). After this addition, the solution was cooled to a temperature between 110 and 130°C. The dots are then allowed to grow for 1-10 minutes at this temperature. Varying the injection parameters and growth temperatures results in a change of the spectral position of the first exciton peak; accordingly, varying these parameters leads to a change in the photoluminescence of the nearly monodisperse colloid of PbSe nanocrystals. As a final steps, the dots were cooled to room temperature, precipitated out of the solution using methanol, and separated by centrifugation and stored in dry condition. The preparation of quantum dots described by Wehrenberg et al. uses relatively unreactive chemicals and can be performed in a hood.

As discussed previously, many of these fluorescent semiconductor nanocrystals have been coupled covalently to biomolecules as a step towards their use in ultrasensitive biological detection applications. Bruches et al.[6] have prepared CdSe-based quantum-dot semiconductor nanocrystals as fluorescent biological labels. CdSe-based quantum dots are of special interest since it was discovered[84] that a thin ZnS capping on a 2.7-to-3.0-nm diameter CdSe quantum dot passivates the core CdSe quantum dot with the result that high quantum yields of 50 % are observed at room temperature. As discussed previously, these nanometer-sized quantum dots are detected through laser-stimulated photoluminescence and biomolecules attached to the quantum dots are used for purposes such as recognition of specific analytes including proteins, DNA, and viruses. These flourescent nanocrystals were found to have a narrow, tunable, symmetric emission spectrum as well as to be photochemically stable. In these studies'[6] core-shell particles of CdSe-CdS were enclosed in silica shells in order to make them soluble in water. The utility of these nanocrystals as biotags was demonstrated by using them to fluorescently label mouse fibroblast cells with silica-coated CdSe-CdS nanocrystals of two different diameters so that they would fluoresce at two wavelengths: the larger 4-nm-core nanocrystals emitted red radiation with a spectral maximum at 630 nm and a quantum yield of 6 %, and the smaller 2-nm-core nanocrystals emitted green radiation with a spectral maximum at 550 nm and a quantum yield of 15 %. The surfaces of these quantum dots were modified to interact selectively with the biological sample by (a) specific ligand-receptor interactions, or (b) electrostatic and hydrogen-bonding interactions. In case (a), the avidin-biotin interaction was used to specifically label the F-actin filaments with the 4-nm-core red-emitting nanocrystals. In case (b), nanocrystals coated with trimethoxysilylpropyl urea and acetate groups were observed to bind with high affinity in the cell nucleus. In this former case, biotin was bound covalently to the nanocrystals and these nanocrystals were then used to label fibroblasts that had been incubated in streptavidin and phalloidin-biotin. Imaging these samples was done with conventional wide-field microscopes as well as with laser-scanning-confocal-fluorescence.

Chan and Nie[7] have used highly luminescent ZnS-capped CdSe quantum dots covalently coupled to biomolecules for ultrasensitive biological detection. In particular, nanometer-sized quantum dots were detected through laser-stimulated photoluminescence and biomolecules attached to the quantum dots were used to recognize analytes. Chan and Nie[7] used mercaptoacetic acid for solubilization as well as for covalent protein attachment. It was found that the mercapto group binds to a Zn atom when it reacts with the ZnS-capped CdSe quantum dots in chloroform, the polar carboxylic acid group renders the quantum dots water soluble, and that the free carboxyl group is available for covalent coupling to biomolecules by cross-linking to reactive amine groups. In the case of covalently attached proteins, Chan and Nie[7], demonstrated that the ZnS-capped CdSe quantum dots were biocompatible in living cells. Chan and Nie acquired fluorescent images from cultured HeLa cells that had been incubated with control samples of mercapto-quantum-dots and with tranferrin-quantum-dot conjugates. In the absence of transferrin, no quantum dots were observed inside the cell. Chan and Nie[7] also investigated the use of quantum dot labels for sensitive immunoassays. In particular, fluorescent images were obtained of quantum-dot-immunoglobulin G (IgG) conjugates that were incubated with with a specific polyclonal antibody (0.5 μg/ml) and bovine serum albumin (BSA) (0.5 mg/ml). It was observed that the polyclonal antibody recognized the Fab fragments of the immunoglobulin and led to extensive aggregation of the quantum dots. In the case of BSA, well-dispersed, primarily single-quantum-dots were observed.

CdSe-based nanocrystals have also been studies for their utility as DNA-functionalized biotgas. In particular, Mitchell, Mirkin, and Letsinger[10] have developed techniques for producing DNA-functionalized CdSe-based quantum dots. This development is especially significant since DNA has exceptional binding specificity. Mitchell, Mirkin, and Letsinger[10] have used 3-mercaptopropionoc acid to passivate the quantum dot surface and to act as a pH trigger for controlling water solubility and subsequent oligonucleotide surface immobilization. In this study,[10] an excess of 0.10 mL of 3-mercaptopropionoc acid was used to react with a suspension of CdSe/ZnS quantum dots --- coated with a mixture of trioctylphosphine oxide(TOPO)/trioctylphosphine (TOP) --- in 1.0 mL of N,N-dimethylformamide to form propionic acid functionalized quantum dots. As is well known, TOPO/TOP-coated quantum dots are soluble only in nonpolar solvents. For these quantum dots, the presence of surface bound propionic acid was indicated by the characteristic ν_{CO} band. These quantum dots are essentially insoluble in water but their solubility is enhanced by deprotonating the surface bound mercaptopropionic acid with 4-(dimethylamino)-pyridine. The resulting quantum dots that were readily soluable in water and were stable for up to a week. Mitchell, Mirkin, and Letsinger[10] also reported the first successful modification of semiconductor nanocrystals with single-stranded DNA, the generation of DNA-linked quantum dot assemblies, and a preliminary study of the optical properties of these structures.

Mattoussi et al.[80] have demonstrated the self-assembly of CdSe-ZnS quantum dot bioconjugates using engineered recombinant proteins. These protein-molecule-conjugated luminescent CdSe/ZnS core-shell nanocrystals have applications as bioactive fluorescent probes in imaging and sensing as well as in immunoassay. Mattoussi et al.[80] used a chimeric fusion protein that binds electrostatically to the oppositely charged surface of the capped quantum dots, and they developed a conjugation method based on self-assembly utilizing electrostatic attractions between negatively-charged lipoic acid capped CdSe-ZnS quantum dots and engineered bifunctional recombinant proteins consisting of positively charged entities --- containing a leucine zipper --- genetically fused with desired biologically relevant molecules. Water-soluble CdSe-ZnS core-shell nanoparticles were prepared by Mattoussi et al.[80] as follows: TOPO/TOP capping groups were exchanged with dihydrolipoic acid groups by suspending 100-300 mg of TOPO/TOP-capped dots after size selection precipitation in 150-500 μL of dihydrolipoic acid; after dilution with about 1.5 mL of dimethylformamide, deprotonation of the terminal lipoic acid –COOH groups was carried out by adding potassium tert-butoxide; centrifugation was used to form a sediment of the resulting precipitate of nanoparticles and released TOPO/TOP reagents; and the sediment was dispersed in water and centrifugation/filtration was used to remove TOPO/TOP resulting in a clear dispersion of alkyl-COOH capped nanocrystals. The emission characteristics for these stable aqueous quantum-dot dispersions were found to be the same as those of the initial nanoparticles; namely, they exhibited a photoluminescent yield of about 10-20 %. These lipoic-acid-capped core-shell nanoparticles were then conjugated with maltose-binding-protein-basic leucine zipper (MBP-zb) protein in 5 mM sodium borate at pH 9.

Rosenthal et al.[117] have used serotonin-labeled fluorescent CdSe nanocrystrals (SNACs) to interact with Drosophila serotonin (dSERT) and human serotonin (hSERT) transponders expressed in both HeLa cells and human epithelial kidney cells (HEK-293 cells) in vitro. In this work, Rosenthal et al. synthesized SNACs as follows: (a) 60 mg of serotonin was reacted with 1 mL of a 20% tetra methyl ammonium hydroxide/methanol solution in 10 mL of methanol for 30 min. under nitrogen at room temperature; (b) 30 mg of 30 Å trioctylphosphine-oxide-coated (TOPO-coated) CdSe nanocrystals were then added to produce a clear red solution; (c) the reaction mixture was reduced to 3 mL under vacuum to isolate the SNACs; (d) SNACs were precipated with 10 mL of acetone; (e) the solution was then further redissolved in 3 mL of methanol and again precipitated with 10 mL of acetone; and (f) the concentration of the SNAC solution was determined by UV-visible spectroscopy. In this work, Rosenthal et al.[117] synthesized a serotonin-linker arm ligand (1-[3-(2-amino ethyl)-1H-indil-5-yloxy]-3,6-dioxa-8-mercaptooctane). Specifically, an N-protected derivatives of serotonin was used by Rosenthal et al. to synthesize (1-[3-(2-Amino ethyl)-1H-indil-5-yloxy]-3, 6-dioxa-8-mercaptooctane). The hydroxyl group of the N-protected derivative of serotonin was coupled with a linker arm, which contained a thiol group. First the protecting group from the serotonin derivative was removed and then the protecting group on thiol was removed. This resulted in (1-[3-(2-Amino ethyl)-1H-indil-5-yloxy]-3, 6-dioxa-8-mercaptooctane). This process employed serotonin protected by using a phthalimido group to give a N,N-phthalimido-2-(5-hydroxy-1H-indole-3yl)ethylamine. One end of the generic linker arm has a p-methoxy benzyl thio ether and the other end consisted of a poly (ethylene glycol) derivative with a tosylate. In this work,[117] the derivative is based on the use tri ethylene glycol for the synthesis. Sodium salt of p-metoxylbenzylthiol replaces the chlorine atom by refluxing 2-[2-(2-chloroethoxy)ethoxy]ethanol in ethanol for a period of 24 hours in nitrogen environment. A 71% yield of 8-(4-methoxybenlylthio)-3,6-dioxaoctanol was obtained by nucleophilic displacement of chlorine. 8(4-methoxybenzylthiol)-3,6-dioxaoctanol was stirred in pyridine with an excess of tosyl chloride resulting in a tosylate. The yield of 8(4-methoxybenzylthiol)-3,6-dioxaoctanol tosylate realized by this procedure was 72%. By refluxing in acetone, N,N-phthalimido-2-(5-hydroxy-1H-indole-3yl)ethylamine was coupled to the linker arm; this procedure was carried out in the presence of 3 eq of cesium carbonate for 18 hours resulting in a 70% yield of 1-[3-[2-(N,N-phthalimido)ethyl]-1H-indol-5-yloxyl]-3,6-dioxa-8-(4-methoxybenzylthio) octane. By stirring 1-[3-[2-(N,N-phthalimido)ethyl]-1H-indol-5-yloxyl]-3,6-dioxa-8-(4-methoxybenzylthio) octane for 2 hours at room temperature in ethanol in the presence of excess hydrazine hydrate removed the phthalimido functionality giving a yield of 51% of 1-[3-(2-aminoethyl)-1H-indol-5-yloxy]-3,6-dioxa-8-(4-methoxybenzylthio)octane. 1-[3-(2-aminoethyl)-1H-indol-5-yloxy]-3,6-dioxa-8-(4-methoxybenzylthio)octane was stirred in trifluoroacetic acid at 0 ^{0}C to remove the p-methoxybenzyl protecting group on the sulfur atom. This is then

further treated with hydrogen sulphide at room temperature in glacial acetic acid. A yield of 39% of 1-[3-(2-aminoethyl)-1H-indol-5-yloxy]-3,6-dioxy-8-mercaptooctane was obtained. The synthesis of the serotonin-linker arm – nanocrystal (core-shell nanocrystal) conjugates was accomplished by Rosenthal et al. as follows: (a) 75-Å-diameter TOPO-coated CdSe/ZnS core/shell nanocrystals were synthesized from 30 Å cores; (b) the TOPO ligands were exchanged with pyridine at 60 °C in order to attach the serotonin ligand to the core/shells; (c) dichloromethane was used to dissolve the serotonin ligands and they was later added to a pyridine solution maintained at 60 °C; (d) the nanocrystals were cooled to room temperature after and hexanes were then used to precipitate the nanocrystals;
(e) mercaptoacetic acid was added to the surface of nanocrystals to inhibit the stearic interactions between the ligands as also to prevent the interference of ligand-SERT interaction; (f) equal volumes of DMF (1 mL) and mercaptoacetic acid (1 mL) were used to dissolve to serotonin-linker-arm-conjugated nanocrystals (LSNACs) and this mixture was stirred for 24 hours at room temperature; (g) the mercaptoacetic acid was neutralized adding 2 M equivalent of potassium tert-butoxide; (h) this was followed by centrifugation, washing with methanol (20 mL) and drying under reduced pressure to obtain the desired serotonin-linker-arm-conjugated nanocrystals (LSNACs). Rosenthal et al. then performed electrophysiological measurements showing that the LSNACs produced currents when exposed to serotonin-3 ($5HT_3$) transporter did not result in currents from the serotonin-3 ($5HT_3$) receptor. In addition, it was found that SERT-transfected cells were labeled by these LSNACs.

In a study by Akerman et al.,[118] tri-n-octylphosphine oxide-coated ZnS-capped CdSe quantum dots were synthesized and coated with mercaptoacetic acid to render them water-soluble. To study the binding of these quantum-dot-based to specific biological structures, these quantum dots were coated with three peptides: (a) CGFECVRQCPERC peptide (GFE peptide), which in lung blood vessels binds to the membrane dipeptidase on the endothelial cells in the lung blood vessels: (b) KDEPQRRSARLSAKPAPPKPEPKPKKAPAKK peptide (F3 peptide) which in various tumors preferentially binds to the blood vessels and tumor cells; and (c) CGNKRTRGC (LyP-1) which --- in certain tumors --- recognizes the lymphatic vessels and the tumor cells.

Peptide synthesization was carried out by N-(9-fluorenylmethoxycarbonyl)-L-amino acids chemistry with a solid-phase synthesizer. In addition, 3-mercaptopropionimidate hydrochloride --- an imidoester compound which contains a sulfohydryl group --- was used for the thiolation of the peptides. Iminothiolane was employed for incubation of the peptides for one hour in 10 mM PBS at a pH of 7.4 at a 1:1 molar ratio. The mercaptoacetic acid coated quantum dots were then added to the peptide-containing solution in order to replace some of the mercaptoacetic acid groups with the thiolated peptide. Coadsorption of polyethylene glycol and peptides was achieved by thiolation of amine-terminated PEG with iminothiolane. A solution of mercaptoacetic acid coated quantum dots in 10 mM PBS maintained at pH 7.4 was used for addition of thiolated PEG. The thiolated peptide was added to the PEG/QD solution at room temperature for overnight incubation. Purification of the coated quantum dots was carried out in micro spin G-50 columns. In these studies, these peptide-functionalized QDs were injected (as a solution of 100 – 200 μg in 0.1 – 0.2 mL PBS) into the tail of a mouse and allowed to circulate for periods of 5 minutes or more. By extracting the indicated tissues and by observing the luminescence from the peptide-functionalized quantum dots, it was established that the functionalized QDs have successfully labeled lung blood vessels, tumor cells, and lymphatic vessels. These studies provide direct evidence for the in vivo use of QD biotags.

Wu et. al.[5] have made major demonstrations of the utility of using semiconductor quantum dots as fluorescent biotags of nuclear antigens inside cell nuclei, actin and microtubules, and of the breast cancer marker Her2 on the surfaces of both fixed and live cancer cells. These biotags were CdSe ZnS-coated semiconductor quantum dots functionalized with streptavidin and IgG antibody (Quantum Dot Corporation). These studies demonstrate that semiconductor quantum dots are a powerful tool in the study of subcellular phenomena with nanoscale precision. Wu et al. isolated CdSe-ZnS quantum dots from hexanes and ligand solution with an equal volume of methanol. This was rinsed with methanol and further redispersed in $CHCl_3$. Neutralized amphiphilic polymer (40% octylamine-modified polyacrylic acid, 2,000 units/QD) was mixed in this solution followed by solvent evaporation. Wu et al. then redispersed the resulting dry film in water and used gel filteration to remove excess polymer. The nanocrystal surfaces were then cross-linked using the frequently-used cross-linking process based on EDC (1-ethyl-3-(3-dimethylamino propyl) carbodiimide) to achieve the functionalization with antibodies and streptavidin. In the work of Wu et al., nuclear antigens in the nuclei of human epithelial cell were labeled with anti-nuclear antigen, anti-human IgG-biotin, and CdSe-ZnS-streptavidin quantum dots emitting in the red. Moreover, the nuclei of 3T3 cells were stained with anti-nuclear antigen, anti-human IgG-biotin, and CdSe-ZnS-streptavidin quantum dots emitting in the red. The labeling of microtubules was accomplished with anti-a-tubulin antibody, anti-mouse IgG biotin, and CdSe-ZnS-streptavidin quantum dots emitting in the green. In addition, Wu et al. demonstrated that (a) Her2 on the surface of SK-BR-3 cells could be imaged using mouse anti-Her2 antibody and CdSe-ZnS-streptavidin quantum dots emitting in the green and that (b) nuclear antigens could be imaged using anti-nuclear antigen, anti-human IgG-biotin, and CdSe-ZnS-streptavidin quantum dots emitting in the red. Hence, the simultaneous detection of two-color luminescence was demonstrated for a variety of conditions, providing yet another indication of the great utility of quantum dots as biotags.

Jaiswal et al.[4] have used the technique of electrostatic self-assembly of negatively charged dihydrolipoic-acid-capped CdSe-ZnS quantum dots with positively charged proteins to accomplish long-term multiple-color imaging of live cells for periods of over a week as the cells developed and grew. This study was based in part on the earlier work of Mattoussi et al.[119] These techniques accommodate the use of naturally charged molecules such as avidin as bridging proteins, but they are also amenable to the use of a general protein which is fused to a positively-charged basic leucine-zipper peptide. These techniques provide a flexible means of binding any desired antibody to any dihydrolipoic-acid-capped (DHLA-capped) quantum dot (QD). Accordingly, in the studies of Jaiswal et al., a synthetically engineered protein G-zb (leucine zipper-containing peptide fused to the B2 binding domain of streptococcal protein G) or avidin was used to conjugate antibodies to colloidal quantum dots. The noninvasive labeling of mammalian HeLa cells through endocytosis of DHLA-capped quantum dots was observed. These studies provide evidence that the DHLA-capped QDs did not interfere with normal cellular functions such as endocytisis, motility, and cellular signaling. In fact, Jaiswal et al. conclude that these approaches for non-invasive cell labeling are viable for periods of over 12 days and that cell growth and motility are not affected. These findings point to many future uses of QDs in studies of internal subcellular processes.

Chan and Nie[121] have reported on the use of CdSe-ZnS quantum dots (QDs) functionalized with selected biomolecules for biological detection. In the studies reported by Chan and Nie, CdSe-ZnS QDs labeled with transferrin proteins were observed to undergo receptor mediated endocytosis in HeLa cells. Moreover, QDs labeled with selected immunomolecules recognized specific antigens and antibodies. Chan and Nie reported that the QD complexes used in their studies were 20 times brighter than organic dyes such as rhodamine, and 100 times more stable against photobleaching.

Dahan et al.[122] have used streptavidin-functionalied CdSe-ZnS quantum dots (SQDs) --- obtained form Quantum Dot Corporation --- to track and image glycine receptor (GlyR) diffusion dynamics in a neuronal membrane for time scales varying from microseconds to minutes. GlyR dynamics was studied since GlyR is the main inhibitory neurotransmitter receptor in the adult spinal cord. The scaffolding protein gephyrin stabilizes GlyR clusters. In this study, the detection of endogeneous GlyR $\alpha 1$ subunits at spinal cultured neuron surfaces was realized via the use of mAb2b primary antibody, biotinilated anti-mouse Fab fragments, and SQDs emitting at 605 nm. In these experiments, spinal cord neurons of E14 Sprague-Dawley rats (1.25×10^5 cells/mL) were incubated with mAb2b primary antibody (2.5 μg/mL) and subsequently incubated with biotinylated anti-mouse Fab antibody (2.8 – 9 5 μg/mL, Fab/Biotin ratio of about 1:0.8). Exposure to of these biotinylated neurons to SQDs was accomplished by using coverslips that had been incubated in streptavidin-functionalized CdSe-ZnS quantum dots (0.2 – 0.7 nM) in a sucrose-supplemented borate buffer (50 mM). The SQD-GlyR complexes were then imaged at room temperature with an inverted microscope equipped with a 60X objective with a N.A. of 1.45. By following the trajectories of SQD-GlyR complexes emitting at 605 nm, diffusion coefficients of the order-of-magnitude of 0.1 μm^2/sec. As a means of comparing labeling with quantum dots to labeling with conventional dyes, Cy3 dye was directly coupled to mAb2B and Cy3 detection was accomplished with a frequency-doubled YAG laser operating at 532 nm. It was found that the trajectories of SQD-labeled GlyRs could be visualized for at least twenty minutes whereas those labeled with Cy3 were visible for about 5 seconds. The signal-to-noise ratio (50 for a 75 ms

Compound Semiconductor	Functionalization	Application
CdSe-based quantum-dot semiconductor nanocrystals: fluorescent biological labels	Trimethoxysilylpropyl urea & acetate groups[6]	Binds with high affinity to cell nucleus due to electrostatic and hydrogen-bonding interactions
CdSe-based quantum-dot semiconductor nanocrystals: fluorescent biological labels	Avidin for avidin-biotin interaction[6]	Specifically label the F-actin filaments
CdSe-based quantum-dot semiconductor nanocrystals: fluorescent biological labels	Serotonin-linker-arm[117]	Binding to human serotonin transponder expresses in both HeLa cells and human epithelial kidney cells
CdSe-based quantum-dot semiconductor nanocrystals: fluorescent biological labels	Mercaptoacetic acid (mercapto group binds to Zn) (polar carboxylic acid renders QD water soluable) (free carboxyl group is available for covalent coupling to biomolecules)[7]	Coupling to biomolecules (by cross-linking to reactive amine groups) --- - nucleic acids - peptides - proteins
CdSe-based quantum-dot semiconductor nanocrystals: coated with trioctylphosphine oxide (TOPO) / trioctylphosphine (TOP)	Excess of mercaptopropionic acid in solution of dimethylformamide with QDs in suspension to form propionic functionalized QDs[10]	Binding to single-stranded DNA Generation of DNA-linked QD assemblies
CdSe-based quantum-dot semiconductor nanocrystals: coated with trioctylphoshhine oxide · coated with mercaptoacetic acid	CGFECVRQCPERC[118] CGNKRTRGC[118]	Binds to endothelial cells in the lung blood vessels Binds to lymphatic vessels of selected tumors

Table 2. Representative functionalization schemes as well as intended applications.

Compound Semiconductor	Functionalization	Application
CdSe-based quantum-dot semiconductor nanocrystals: coated with trioctylphoshhine oxide (TOPO) / trioctylphosphine (TOP)	Chimeric fusion protein binds electrostatically to oppositely charged surface of QD Here the conjugation method is based on the electrostatic self-assembly of: negatively-charged lipoic acid capped CdSe-ZnS QDs and engineered bifunctional[80] recombinant proteins consisting of positively charged entities --- containing a leucine zipper -- - genetically fused with desired biologically relevant molecules	Binding DNA
CdSe ZnS-coated quantum dots	Streptavidin[5] IgG[5]	Binding to cellular features labeled with anti-IgG-biotin
CdSe ZnS-coated quantum dots	Streptavidin[122]	Binds to (as an example) biotinylated neurons

Table 3. Representative functionalization schemes as well as intended applications.

integration time) for the SQDs was an order of magnitude larger than that for the dye. Moreover, the lateral resolution with SQDs was about 5 – 10 nm and with dyes was about 40 nm. Tables 2 and 3 summarize representative functionalization schemes as well as intended applications.

3. RELEVANT PHYSICAL PROPERTIES OF SEMICONDUCTOR QUANTUM DOTS

It is recognized widely in the semiconductor device community that normal modes of lattice vibrations --- known as phonons lead to changes in the optical and electronic properties of semiconductors. Regarding the high-frequency optical phonons, Klein et al.[94] have derived expressions for the longitudinal-optical (LO) and surface-optical phonon modes in CdSe nanospheres and have modeled the size dependence of electron-phonon coupling in these semiconductor nanocrystals. Furthermore, by comparing theory with experimental results, Klein et al.[94] confirm the size independence of the predicted electron-phonon coupling constant and they establish the existence of surface-optical (SO) modes. SO phonon modes have been studied by a number of authors for CdSe quantum dots as well as for a variety of other semiconductor quantum dots.[123-132] In this account, the role of the low-frequency acoustic modes are highlighted since they influence the linewidth of the photoluminescence.

The electronic, optical, and mechanical properties of semiconductor nanocrystals have been investigated by a broad community of scientists and engineers for over a decade.[85-141] One of the dominant physical characteristics of a pure, defect-free semiconductor is that it exhibits a gap in energy, E_{gap}, where no states exist. Above this energy gap is a band of energy states known as the conduction band and below the energy gap is a band of energy states known as the valence band. In the absence of thermal excitation of electrons from the valence band to the conduction band, the conduction band is empty and the valence band is full. Traditionally, the lowest energy in the conduction band of an idealized one-dimensional semiconductor is denoted by $E_c(x)$ and the highest energy in the valence band by $E_v(x)$. In the absence of electric-field or dimensional-confinement effects, $E_c(x)$ and $E_v(x)$ do not depend on x; that is, $E_c(x) = E_c$ = a constant and $E_v(x) = E_v$ = a constant.

In the presence of an electric field, the electrons (and holes in the general case) gain energy. Since the kinetic energy of the electrons in the conduction band may be measured relative to the bottom of the band, the presence of a constant electric field, E, requires that $dE_c(x)/dx = qE$. As a result of this "band-bending" relation, the electrons gain energy due to E as reflected by the fact that $E_c(x)$ (the energy level corresponding to zero kinetic energy) has a slope, $dE_c(x)/dx$ = constant when the electric field is constant. Thus, for a negatively-directed electric field $E = -n_x \mid E \mid$ --- where n_x is a unit vector in the x-direction --- the electron is accelerated to the right and the negative quantity $dE_c(x)/dx$ leads to increasing electron kinetic energy as the electron moves to the right.

Consider now the case where there is no electric field but where the semiconductor has a finite extent, d, in the x-direction. At the boundaries of the semiconductor, the energy states must change to account for the fact that there is a transition between the semiconductor and the surrounding medium. This surrounding medium could be a free-space, another semiconductor, a biomolecule, water, hexane, etc. In many situations, it turns out that the role of the boundaries may be taken into account approximately by taking E_c = constant in the region of the semiconductor, $0 < x < d$, and by taking E_c to have infinite value outside the region $0 < x < d$. Recalling that it is convenient to take E_c as the energy where the kinetic energy of the particle is zero, it follows that electrons with finite kinetic energies may exist in the region between $0 < x < d$ but not outside of $0 < x < d$. For this reason, a semiconductor of finite and small length, d, is known as a quantum well. Indeed, the electrons are confined to this well since states are available inside the well but not in the surrounding region. In the case of a quantum dot, the confinement occurs in three dimensions, and not in one dimension as for the quantum well. In the case of a one-dimensional infinitely-deep (E_c has infinite value outside the region $0 < x < d$ and a zero value inside the well) quantum well, quantum mechanics tells us that the electron wavelengths, λ_n, must fit into the confinement region of width, d, in such a way that the wave amplitudes vanish at $x = 0$ and $x = d$. Thus, $\lambda_n/2 = d/n$ where n = 1, 2, 3, … . Quantum mechanics also tells us that the wavelength of the electron is related to the momentum of the electron, p, through $\lambda = h/p$ where h is Planck's constant with a value of 6.6×10^{-34} Joule-second. This last relationship is known as the de Broglie relation and λ is known as the de Broglie wavelength of an electron of momentum, $p = mv$. Since the kinetic energy, K.E., of the electron is K.E. $= mv^2/2 = p^2/2m$, it follows that,

$$\text{K.E.} = h^2/2m\lambda^2 = n^2h^2/2m(2d)^2 = n^2h^2/8md^2.$$

Thus, the kinetic energies of the electrons in a one-dimensional quantum well of width, d, must take on the specific values, $n^2h^2/8md^2$ where n is known as the quantum number. The kinetic energy for n = 1 is known as the ground state energy; obviously, this state has the lowest kinetic energy of any state in the quantum well. For a spherical quantum well, or quantum dot, it turns out that a subset of the allowed kinetic energies is given by simply replacing the width of the one-dimensional well, d, with the radius of the quantum dot, a. In particular, the ground state is given by $h^2/2m_ea^2$, where m_e is the effective mass of the electron in the semiconductor in question. Likewise, the so-called ground state of the charge carriers in the valence band is $h^2/2m_ha^2$, where m_h is the effective mass of the hole. A "hole" in the valence band results when an electron is removed from one of the valence band states in the semiconductor. In general, this hole moves through the semiconductor lattice with an effective mass, m_h, that is different from the electron effective mass, m_e.

For the quantum dots of interest as luminescent biotags, an electron in a full valence band may be "excited" or elevated from the valence band to the conduction band by a photon of energy $h\nu$ where h is Planck's constant and ν is the frequency of the photon, or equivalently, the frequency of the light. The lowest energy, $h\nu$, possible for this photon is,

$$h\nu = h^2/2m_ea^2 + h^2/2m_ha^2 + E_{gap}$$

where the first two terms on the right side of this equation represent the ground state energies of the electron and the hole in the quantum dot of radius, a, and E_{gap} is the energy difference between the conduction and valence band edges, $E_c - E_v$.

As an example of the formalism just developed, the infinite-barrier approximation when applied to ZnS-coated CdSe provides a simple means of estimating the photon frequency, ν, required to excite an electron from the valence band to the conduction band in the CdSe-ZnS core-shell system. The success of this approach is due, in part, to the facts that E_c for ZnS is greater than that for CdSe and E_v for ZnS is greater than that for CdSe. From Table 2, it is apparent that the ZnS-coated CdSe core-shell system has been successfully applied in biological applications. This success is due to the excellent size control that may be achieved for CdSe nanocrystals as well as the fact that a thin (0.6 ± 0.3 nm) ZnS capping on a typically 2.7-to-3.0-nm-diameter CdSe quantum dot passivates the core CdSe nanocrystal thereby reducing the number of surface traps and resulting in a 50 % quantum yield[84] at room temperature. This quantum yield is exceptionally high and it implies that CdSe-ZnS quantum-dot biotags will provide an especially bright luminescent signal.

In addition to the role played by acoustic phonons in determining properties of the photoluminescent spectra of semiconductor quantum dots, the properties of quantum dots are also influenced by dimensional confinement and by their interactions in aqueous solutions. Moreover, the dielectric properties of the quantum dot and of the surrounding materials play important roles in determining the separations between energy levels in the quantum dots. The photoluminescent spectra and binding energies of quantum dots are influenced by their incorporation in polar and non-polar semiconductor, polymer, and aqueous environments.[54, 133-140] As described previously, the optical linewidths of quantum dots are also influenced by phonon-assisted processes that depend on the spherical nature of the vibrational modes in quantum dots.[141] Coatings have played a major role in the development of quantum dot technology. Indeed, coatings have been developed to ensure that specific functionalized quantum dots will be "soluble" in aqueous environments. Silica[6] and polymer coatings[62] have proven effective in this regard.

Regarding the optical properties of nanocrystals, Shiang et al.[74] have used resonant Raman studies of CdS nanocrystals show that nanocrystals of smaller size (less than 70 Angstroms) have decreased strength of electron-phonon and hole-phonon coupling. In these studies, several different sample environments were used and different surface capping molecules were used; in particular, thioglycolate, thiophenol, polyphosphate and glass matrix can be used as the surface capping molecules.

Semiconductor nanocrystals created in electrolytic solutions and semiconductor nanocrystals in biological environments are subject to screening effects caused by mobile ions in their vicinity. Indeed, any charged semiconductor in an electrolyte will experience the fields originating from the charge-induced redistribution of the ions in the electrolyte. In the case of many of the semiconductors common among the colloidal semiconductor nanocrystals --- II-VI semiconductors with wurtzite structure --- there is an internal polarization oriented along the c-axis of the crystal, known as the spontaneous polarization. This spontaneous polarization, P_s, may be represented in terms of an equivalent distribution of surface and volume charges. For a spherical quantum dot, a constant spontaneous polarization leads to an equivalent charge distribution that is a surface charge given by $|P_s|\cos\theta$, where θ ranges from 0 to π and is the angle measured from the c-axis. Such a distribution of charge will lead to a dipole like distribution and will, of course, be accompanied by bandbending characterized by a linear relationship between the electric field (associated with the polarization) and the derivative of the bandedge. When an electrolyte surrounds such a quantum dot, the mobile ions in the electrolyte will to some degree screen the field associated with the spontaneous polarization. Hence, the presence of the electrolyte will lead to a relaxation of the nominal bandbending and a consequent change in the energy levels of the quantum dot. This will in turn lead to a change in the absorption spectrum and the photoluminescent spectrum of the quantum dot. Electroreflectance at semiconductor-electrolyte interface was studied in the pioneering work of Shaklee et al. in 1965.[142]

The blinking of single ZnS-coated CdSe semiconductor quantum dots has been studied by Kuno et al.[143] using confocal microscopy. Using this confocal microscope the distribution of "on" and "off" times has been determined over a 10^9-fold dynamic range of probabilities and for a nonexponential "off-times" over a range of 10^5. The measured τ_{off} distributions were found to obey a power-law distribution of the form $P(\tau_{off}) \propto 1/\tau_{off}^{1+\alpha}$. This is in contrast to the fluorescent intermittency distribution observed in self-assembled InP quantum dots; indeed, Sugisaki et al.[144] find a τ_{off} distribution that scales exponentially as $\exp(-E/k_BT)$ where E is an activation energy associated with trap states influencing the intermittency of blinking, k_B is the Boltzmann constant, and T is the temperature. Based on this exponential distribution, the observation that blinking self-assembled dots are found in the vicinity of scratches, and the observation that the blinking frequency is enhanced dramatically in the presence of near-infrared radiation, Sugisaki et al. have interpreted blinking in self-assembled quantum dots in terms of the electric field associated with carriers trapped at a deep localized center in the $Ga_{0.5}In_{0.5}P$ matrix. Such self-assembled quantum dots are formed, i.e. grown, on a substrate and this structure plays a major role in the interpretation of Sugisaki et al. In contrast, for the case of colloidal quantum dots, Kuno et al. have interpreted that $1/\tau_{off}^{1+\alpha}$ as being due to a distribution of energies of the trap states that interfere with the nominal photoluminescence of a semiconductor. As noted by Kuno et al., such a power-law scaling may also be explained in terms of a distribution of tunneling distances between quantum-dot core and interface states. Moreover, Kuno et al. argue that previous association of blinking in colloidal quantum dots with Auger processes is not consistent with their data. The inverse power-law distribution observed by Kuno et al. is consistent with a distribution of trap/interface states. Such distributions have been discussed in term of Levy distributions[145] and were considered in the pioneering work of Scher and Montroll.[146]

Han et al.[8] have demonstrated the key elements of multicolor optical coding by embedding ZnS-coated CdSe quantum dots (QDs) of different diameters in polymeric microbeads. By controlling the ratios of the different QDs embedded in a given microbead, each microbead is endowed with a particular emission spectrum that serves as a "fingerprint" in distinguishing the large number of different microbeads that may be fabricated in this way. For example, five different QDs diameters will result in five distinct colors, and the simultaneous use of ten different intensity levels --- i.e., different numbers of QDs in a given microbead --- will result in 100,000 distinct emission spectra from the ensemble of possible microbeads. The studies of Han et al. indicate that cross-linked beads, formed by emulsion polymerization of styrene, divinylbenzene, and acrylic acid, are appropriate microbead materials for the incorporation of quantum dots. The quantum dots used in these studies are embedded quantum dots are hydrophobic. The microbeads were 1.2 μm in diameter and the results reported by Han et al. indicate that spatial separation of QDs prevents fluorescent resonant energy transfer (FRET) between the QDs embedded in the microbeads. It appears that the bead's porous structure acts as a matrix to spatially separate the embedded QDs. The ability to fabricate a large number of microbeads with distinct spectral features opens the way to new applications in gene expression, medical diagnostics, and high-throughout screening. In addition to the Raman studies of CdS reported by Shiang et al.,[74] Schreder et al.[147] report on a study of confinement effects in CdS quantum dots. Phonon confinement shifts were not observed by these researchers but the measured linewidths of the overtone series points to LO phonon decay into acoustic phonons as a dominant relaxation mechanism.

Bernardini et al.[148] have studied the fundamental aspects of the spontaneous polarization found in würtzite III-V nitrides and they model the numerical values of the spontaneous polarization for several cases on interest; specifically, they report for several semiconductors the values AlN (-0.081 C/m^2), GaN (-0.029 C/m^2), InN (-0.032 C/m^2), ZnO (-0.057 C/m^2), and BeO (-0.045 C/m^2). In the next section, these spontaneous polarizations will take on special significance in the application of quantum dots in biological systems.

4. CONCEPTS AND TOOLS UNDERLYING THE INTEGRATION OF QUANTUM DOTS WITH BIOLOGICAL SYSTEMS

In order to develop concepts and tools necessary to integrate nanoscale semiconductor quantum dots with biological structures, it is essential to take into account the physical properties of nanocrystals discussed in the last section. As an example, the spontaneous polarizations of nanocrystals produce electric fields as a result of the equivalent surface charge given by $|P_s|\cos\theta$, where θ ranges from 0 to π and is the angle measured from the c-axis. For CdSe quantum dots,[110, 9] this surface charge may lead to a potential difference across the quantum dot with a magnitude of about a quarter of a volt! A potential difference of this magnitude is indeed significant since Hodgkin and

Huxley[149] showed that the voltage-dependent switching --- opening and closing --- of K and Na ion channels are associated with the action potential[150] and that a 4 mV change[100] in the potential causes an e-fold change in the ratio of the open probability to the closed probability. Taking the associated energy difference of, $V\Delta q$, to be of the order k_BT (about 26 mV at room temperature), it follows that $\Delta q = 6e$ and that about six electrons are needed to switch such an ion channel. The authors thus view the use the photoelectric properties of quantum dots as active electronic interfaces to be difficult to achieve since the Coulomb interaction works strongly against the photoproduction of six electrons from a single quantum dot! However, the authors predict that the fields produced by the spontaneous polarization of a wurtzite quantum dot is sufficient to gate an ion channel if the quantum dot is anchored at the site of the ion channel. Indeed, the potential difference across a quantum dot with a modest spontaneous polarization (as for CdSe or CdS) is significantly larger that the 4 mV needed to "switch" the ion channel. For wurtzites like GaN, AlN, and ZnO the spontaneous polarization are more than an order of magnitude great than those in CdSe and CdS, and the potential difference across the quantum dots increase in proportion to these spontaneous polarizations. It is thus predicted that wurtzite quantum dots may be used to switch such ion channels, thereby controlling cellular functions through the integration of nanoscale semiconductor elements with biological structures. The authors view this to be just one of the many revolutionary ways that integrating nanostructures with biological structures will have an impact on biology and medicine.

As mentioned previously, peptides may be used to direct the binding of a peptide-functionalized quantum dots with integrins present in cellular membranes. Tables 4 and 5 summarize a number of peptides along with the associated integrins participating in the peptide-integrin binding pairs.[151-157] The peptides identified in Table 3 are of course composed of the twenty amino acids; these are: alanine (A), arginine (R), asparagine (N), aspartic acid (D), cysteine (C), glutamic acid (E), glutamine (Q), glycine (G), histidine (H), isoleucine (I), leveine (L), lysine (K), methionine (M), phenylalanine (F), proline (P), serine (S), threonine (T), tyrosine (Y), tryptophan (W), and valine (V). Thus, as an example RGD is the chain of arginine-glycine-aspartic acid. In recent studies of pertide-functionalized quantum dots binding to integrins,[14,158] peptides such as CGGGRGD are used; the cysteine (C) amino acid will provide for binding to semiconductor nanocrystals through the relatively strong thiol bond and the glycine (G) amino acids serve as linking elements. Glycine is a natural choice for the bridging amino acid between the C terminus amino acids because it has a simple side group of a single hydrogen atom that should produce minimal if any unwanted packing effects. For completeness, it is noted that peptide chain molecules are formed by linking a sequence of amino acids by the binding of the NH_2 group of one amino acid to the COOH group of an adjacent amino acid; these groups are depicted in the Figures 1 and 2; the cysteine and glycine amino acids are shown in these figures.

Peptide	Integrin/Comment
KGD	$\alpha_{IIb}\beta_3$ --- Fibrinogen binds to $\alpha_{IIb}\beta_3$ integrins and causes clotting of blood cells; if RGD is bound to $\alpha_{IIb}\beta_3$ integrins on these cells, clotting is inhibited
PECAM	$\alpha_v\beta_3$ --- Ig superfamily adhesion family protein in endothelial cells, platelets, and leukocytes
RGD	$\alpha_v\beta_5$. --- Fibronectin (RGD) is ligand of $\alpha_v\beta_5$
RGD	$\alpha_v\beta_5$. --- Fibronectin (RGD) is ligand of $\alpha_v\beta_1$
KQAGDV	$\alpha_{IIb}\beta_3$ --- Mimics RGD sequence since it binds to $\alpha_{IIb}\beta_3$
LDV	$\alpha_4\beta_1$ integrin binds to a sequence centered around an LDV motif; LDV is relatively specific in its binding to $\alpha_4\beta_1$
LDV	$\alpha_4\beta_7$ integrin binds to a sequence centered around an LDV motif
DGEA	$\alpha_2\beta_1$ attachs cells of lower layer of epidermis to basement membrane
FYFDLR	$\alpha_2\beta_1$ attachs cells of lower layer of epidermis to basement membrane; FYFDLR sequence in type IV collagen is a $\alpha_2\beta_1$ recognition sequence
YGYYGDALR	$\alpha_2\beta_1$ --- YGYYGDALR sequence in laminin is a $\alpha_2\beta_1$ recognition sequence
KRLDGS	$\alpha_M\beta_2$ --- KRLDGS sequence in fibrinogen is a $\alpha_M\beta_2$ recognition sequence

Table 4. Selected peptides and associated integrins based on References 151-157.

Peptide	Integrin/Comment
IDA(PS)	$\alpha_4\beta_1$ --- IDA(PS) in IIICS domain of fibronectin is an $\alpha_4\beta_1$ recognition sequence; IDA(PS) appears to be a variation of LDV
REDV	$\alpha_4\beta_1$ --- REDV in IIICS domain of fibronectin is an $\alpha_4\beta_1$ recognition sequence; REDV may be a homologue of RGD; REDV is relatively specific in its binding to $\alpha_4\beta_1$
RE	$\alpha_5\beta_1$ attachs cells of lower layer of epidermis to basement membrane Dipeptide RE is similar to RGD
RGDS	$\alpha_5\beta_1$ --- Adhesion to most cells. Fibronectin
RGDV	$\alpha_v\beta_3$ --- Adhesion to most cells. Vitronectin
LDV	$\alpha_4\beta_1$ --- LDV is relatively specific in its binding to $\alpha_4\beta_1$ Binds to cancer cells
RGD	$\alpha_3\beta_1$ --- Fibronectin (RGD)
LEDV	$\alpha_4\beta_1$ --- Fibronectin (LEDV)
SIKVAV	Neurite extension
RNIAEIIKDI	Neurite extension
IKVAV, LQVQLSIR, CGFECVRQCPERC, CGNKRTRGC	Neurite binding (and, in some cases, extension)
DGEA	Adhesion to platelets
VTXG	Adhesion to platelets
48 peptides derived from regions of laminin-1 -- as examples, GTNNWWQSPIQN, and LLEFTSYARYIRL	Neurite binding (and, in some cases, extension)

Table 5. Selected peptides and associated integrins based on References 151-157.

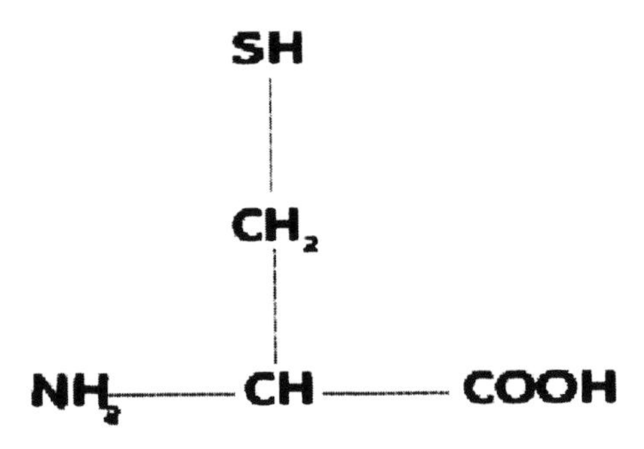

Figure 1. Cysteine amino acid.

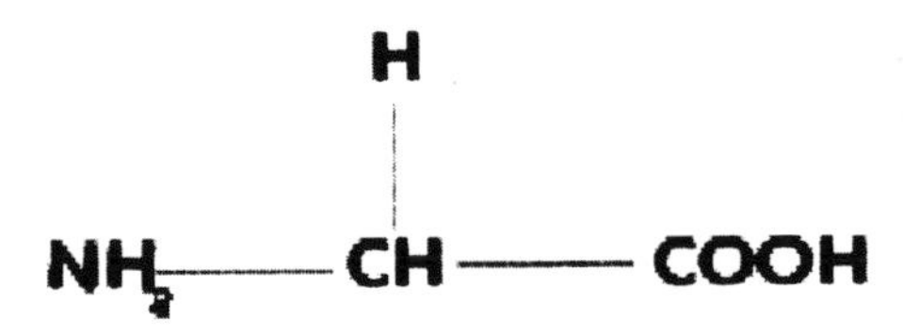

Figure 2. Glycine amino acid.

When a nanostructure is placed in aqueous environments --- as is generally unavoidable when integrating manmade nanostructures with biological structures --- the electronic and optical properties of the nanostructure are changed as a result of the interactions between the nanostructure and its environment. The effects of analogous interactions have been studied for many years in the field of semiconductor microelectronics where dielectric semiconductors are in intimate contact with metals and insulators. For example, in the case of a junction-field-effect transistor, a metal is deposited directly on the electrically-active semiconducting channel of the transistor. Depending on the voltage applied across the metal-semiconductor interface, the electrons available at the interface, and the alignment of the electronic states at the boundary between the metal and the semiconductor,[126] there is a region of variable thickness in the semiconductor --- near the metal-semiconductor interface --- that may be depleted of electrons. Such a depletion region may be micrometers in thickness. For semiconductor quantum dots that have diameters measured on a scale of a few nanometers, it is clearly essential that such boundary related effects be taken into account. As an example of one such effect, a charge placed in a spherical quantum dot will produce an electric field. Associated with the quantum dot and the surrounding material (or materials) are specific --- but confinement dependent --- values of the dielectric constant.[137-140] Generally, bulk materials have a frequency-dependent dielectric constant $\varepsilon(\omega)$ that is independent of the size of the bulk material; however, for dielectric materials with nanoscale dimensional confinement, the energy levels in the material depend on the size of the structure and there is a resulting dependence of the dielectric constant on the size of the nanostructure.[139-140] At the boundary between the quantum dot and the surrounding material the discontinuity in the dielectric constant results in an effective surface charge. Consider the case of a CdSe quantum dot with a 2 nm diameter encased in a spherical shell of ZnS; this CdSe-ZnS core-shell system is then placed in water which has an effective dielectric constant of about 80. The surface charges on each spherical surface will lead to a shift in the potential in the quantum dot that will cause a change in the ground state energy of the single-electron state in the quantum dot.[159] The calculated shift in the one-electron ground-state energy for an electron in such a CdSe quantum dot is shown in Figure 3 for the case where the quantum dot is coated with a ZnS shell.[160] The CdSe-ZnS core-shell quantum dots of Figure 3 are surrounded by a fluid with the dielectric constant of water. The dramatic change in the energy illustrates clearly that it is essential to take into account quantum-confinement and geometrical effects on dielectric screening in determining the electronic properties of nanostructures. A pervasive feature of biological environments is the presence of electrolytic fluids. In these water-based electrolytes, positive and negative ions are mobile and they respond to electric fields. These electrolytes play a fundamental role in determining the membrane

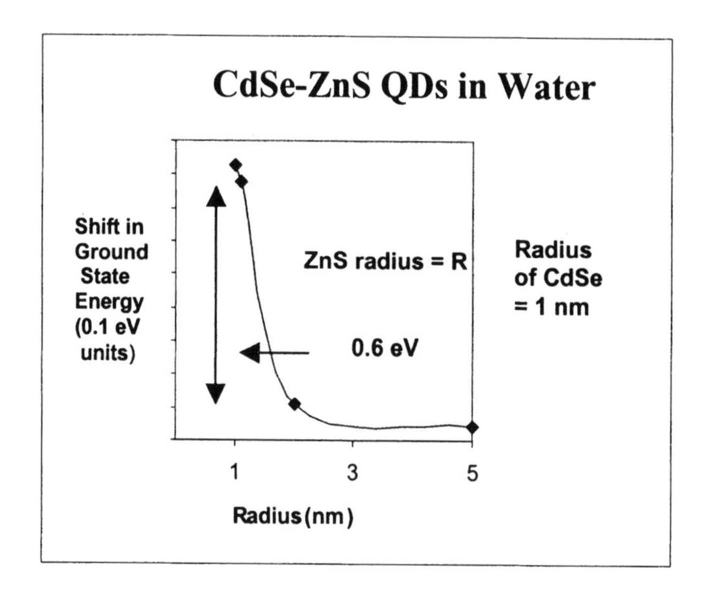

Figure 3. One-electron ground-state energy level (electron volts) for a CdSe-ZnS core-shell quantum dot structure with an outer radius R (nanometers) in an aqueous solution.

potential existing across cellular membranes.[149-150] Indeed, the membrane potential arises as a result of the different concentration of the anions and cations in the electrolyes on the opposite sides of the membrane. As discussed previously, semiconductor quantum dots composed of wurtizes --- including AlN, CdS, CdSe, GaN, ZnO, and ZnS --- manifest intrinsic spontaneous polarization fields. As known from the laws of electrostatics, these fields may be replaced by volume and surface charges generating equivalent fields. As described previously, these charge distributions cause such semiconductor quantum dots to behave as dipoles. Clearly, placing such a dipole in an electrolyte will lead to a rearrangement of the anions and cations in the electrolyte in the vicinity of thenanocrystal. In the limit that the anions and cations are completely free to respond to the dipole field, there will be a field produced in the electrolyte that cancels the dipole field of the nanocrystal. In a similar manner, a nanocrystal with a net positive or negative charge will attract charge of the opposite sign from the electrolyte.

In past applications of quantum dots, the quantum dots are coated with a protective layer such as a polymeric layer or silica that screens the quantum dots from the electrolytic environments found in biological systems. These protective layers render the optical properties of the quantum dots relatively insensitive to the electrolytic environment. However, if the quantum dot or some other nanoscale structure is to be integrated with a biological structure, it is necessary to consider the case where the nanostructure is in direct contact with the biological structure or biological environment. Ramadurai et al.[161] have recently studied the interaction of CdS quantum dots in NaCl electrolytic environments. In particular, the interaction of the ions in a NaCl electrolytic solution with würtzite CdS nanocrystals was studied optically in the case where the nanocrystals are not coated with a protective layer such as a polymeric layer or silica. Ramadurai et al.[161] found that the absorption edge of the CdS nanocrystal shifts by a few percent as the electrolyte concentration is varied over an order of magnitude. These electrolyte-dependent absorption properties suggest that changes in the optical properties of nanocrystals may be used to study the different electrolytic concentrations found in extracellural and intracellular electrolytic concentrations in biological systems. The effects reported for CdS are expected to be even larger for many other würtzite nanocrystals; indeed, GaN and ZnO exhibit spontaneous polarization fields that are an order of magnitude larger than those of CdS.

To estimate the bandbending, $dE_c(x)/dx$, in a nanocrystal caused by the spontaneous polarization, it is possible to use the previously-derived formula, $dE_c(x)/dx = qE$. For a one-dimensional CdS square well, the lowest eigenenergy is approximated by,[162,114]

$$E_{square} = E_{gap} + (h/d)^2/8m_e + (h/d)^2/8m_h - 3.536e^2/(4\pi\varepsilon\varepsilon_o d)$$

where d is the width of the quantum well, h is Planck's constant, the effective mass of the electron, m_e, is taken to be 0.235 of the free electron mass (9.11×10^{-31} kg), the effective mass of the hole, m_h, is taken to be 1.35 of the free hole mass, ε is the relative dielectric constant of CdS – 5.7, ε_o is the permittivity of free space – 8.85×10^{-12} F/m, and e is the electron charge – 1.6×10^{-19} C. The first term in this expression represents the bandgap of bulk CdS, the second and third terms represent the confinement energies, respectively, and the fourth term represents the Coulomb binding energy of the exciton in CdS. This expression is appropriate when there is no bandbending due to the intrinsic spontaneous polarization in CdS. In other words, this expression holds when flat band conditions prevail as would be the case if the surrounding electrolyte were to completely screen the field associated with the spontaneous polarization. As described previously, such screening may be thought of in terms of the anion and cation induced screening of the effective surface charge given by $|P_s| \cos\theta$, where P_s is the spontaneous polarization and θ ranges from 0 to π and is the angle measured from the c-axis. In this case, the nanocrystal behaves as a dipole with positive charge concentrated near one pole, and negative charge concentrated near the other pole. The equatorial plane is, of course, neutral. As discussed previously, the anions and cations in the electrolyte will attempt to screen the spontaneous-polarization-induced dipole. In the case where the electrolyte has such a low density that it does not screen the spontaneous polarization, the ground state energy is given approximately by that of a triangular quantum well:[163]

$$E_{triangle} = E_{gap} + (h^2/8\pi^2 m_e)^{1/3} (9\pi F/8)^{2/3} + (h^2/8\pi^2 m_h)^{1/3} (9\pi F/8)^{2/3} - Fd - 3.536e^2/(4\pi\varepsilon\varepsilon_o d),$$

here F is the electric field associated with the spontaneous polarization. From these results, the corresponding wavelengths, λ_{square} and $\lambda_{trinagle}$ are given by, $\lambda_{square} = hc/E_{square}$ and $\lambda_{trinagle} = hc/ E_{triangle}$. Evaluation of these expressions for CdS results in shift in the absorption edge, and therefore the photoluminescence spectrum, of a few percent consistent with the measurements of Ramadurai et al.[161] The shift in the optical absorption edge as function of electrolytic concentration is yet another effect resulting from the interaction between the nanocrystals and the biological environment.

As demonstrated by Rufo et al.[164,123] a mismatch between the elastic properties of a semiconductor nanocrystal and its surroundings leads to a shift in the acoustic phonon energy; phonon-assisted optical transitions will therefore exhibit phonon sidebands in the photoluminescence spectrum. Figures 4-7 illustrate that the spherical and torsional acoustic modes in CdS and GaN nanocrystals have different frequencies depending on the radius of the nanocrystal as well as on the material surrounding the nanocrystal. Specifically, Figure 4 depicts the energy in meV of spheroidal acoustic phonon mode corresponding to ω_S^{01} as a function of a CdS nanocrystal radius R in nanometers for free-standing, plastic-coated, ZnS-coated, water-encased, and SiO_2-coated CdS nanocrystals. Figure 5 depicts results similar to those of Figure 4 but for the torsional modes. Figure 6 depicts the energy in meV of spheroidal acoustic phonon mode corresponding to ω_S^{01} as a function of a GaN nanocrystal radius R in nanometers for free-standing, plastic-coated,

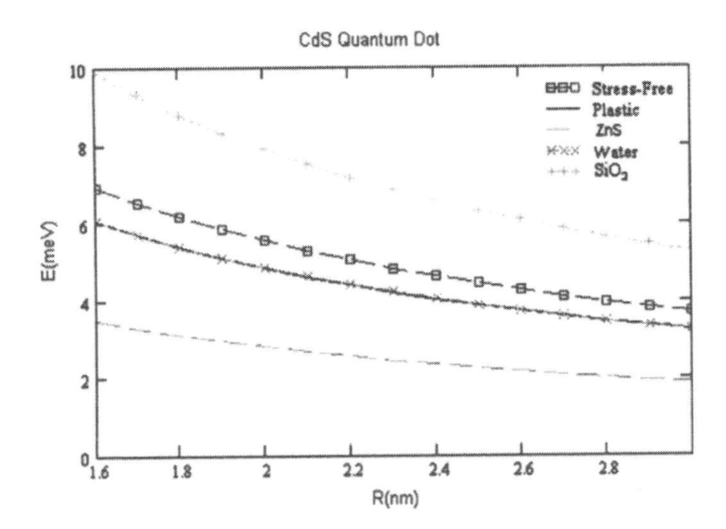

Figure 4. Energy in meV of spheroidal acoustic phonon mode corresponding to ω_S^{01} as a function of quantum dot radius R in nanometers for several cases of interest.

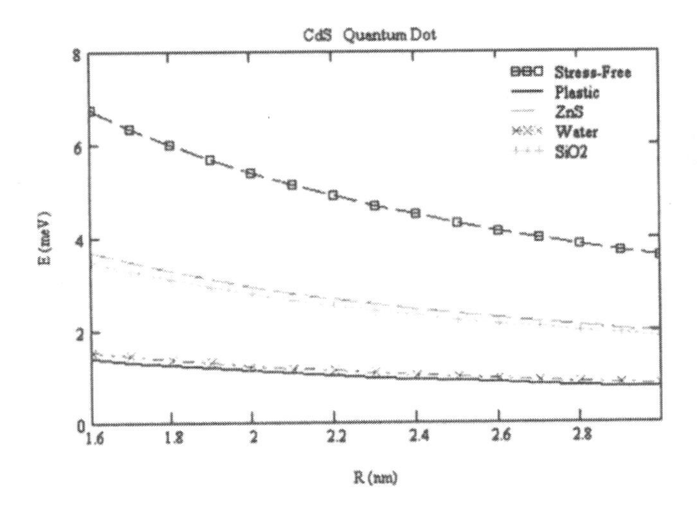

Figure 5. Energy in meV of torsional acoustic phonon mode corresponding to ω_T^{10} as a function of quantum dot radius R in nanometers for several cases of interest.

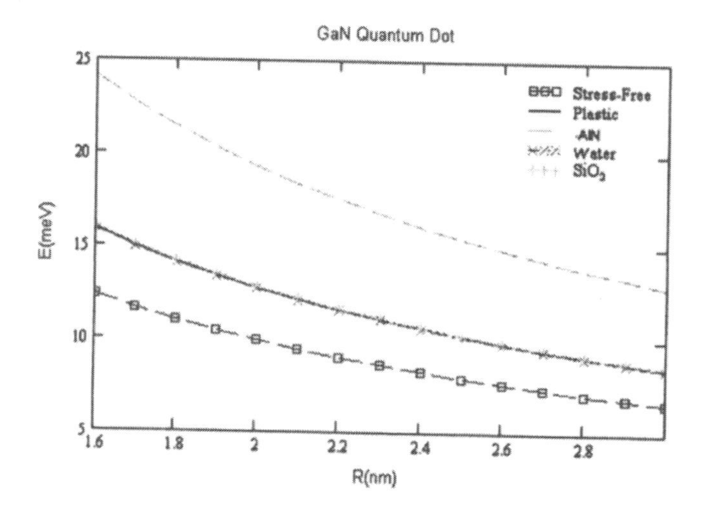

Figure 6. Energy in meV of spheroidal acoustic phonon mode corresponding to ω_S^{01} as a function of quantum dot radius R in nanometers for several cases of interest.

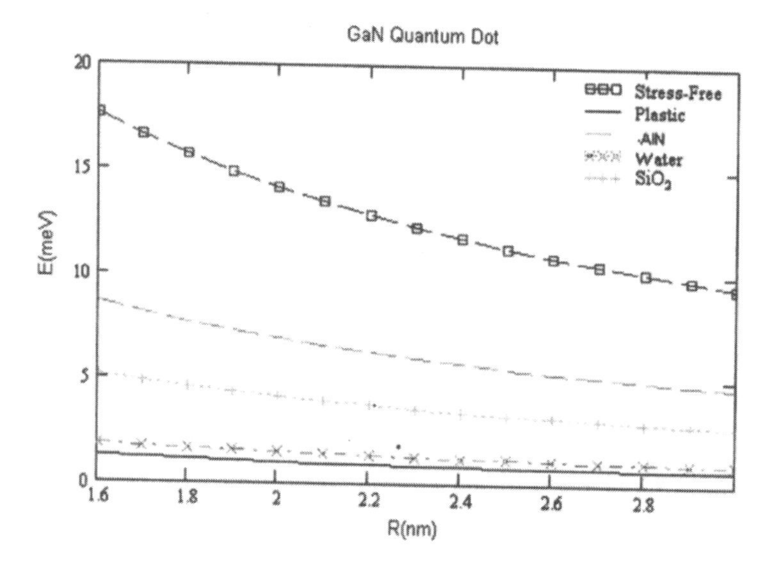

Figure 7. Energy in meV of torsional acoustic phonon mode corresponding to ω_T^{10} as a function of quantum dot radius R in nanometers for several cases of interest.

AlN-coated, water-encased, and SiO_2-coated CdS nanocrystals. Figure 7 depicts results similar to those of Figure 6 but for the torsional modes. From Figures 4-7 it is clear that acoustic-phonon-assisted transitions will lead to sidebands in the photoluminescence spectrum that are shifted by varying amounts depending on the nanocrystal radius as well as the material surrounding the nanocrystal. As an example of special interest in biological applications, it is clear that the presence of water will lead to a particular set of sidebands.

As has been discussed previously, short peptides may be used to facilitate the binding of semiconductor nanocrystals with nanoscale precision to selected cellular integrins.[14, 158] As an example, we have recently bound CGGGLDV to CdS-mercaptoacetic-acid complexes by the thiol bond to the sulfur-containing side group of the cysteine amino acid; in preparing these complexes, 1.8 mg of CGGGLDV was added to 5 mL of the colloidal CdS suspension. The nanocrystal suspension was prepared as described previously.[158] These complexes were then bound by the bond between LDV and integins on fibroblast cells derived from the human fibrosarcoma cell line, HT-1080. In binding the CdS-peptide complex to the fibroblast cells, the cells were exposed to the CdS-peptide suspension for 10 minutes and the cells were then washed five times with a PBS solution. Figures 8a. and 8b. depict white light and fluorescence images of these images fibroblast cells, respectively. The CdS nanocrystals of these studies had diameters of about 5.4 nm. In the case of the fluorescence image, an excitation wavelength of 360 nm was used and the emitted photoluminescent signal was collected for wavelengths above 400 nm; most of the collected radiation was due to red light emitted by surface states on the CdS-peptide complexes. As indicated by the fluorescence pattern in Figure 8b., the integrins appear to be located around the perimeter of the fibroblast cells.

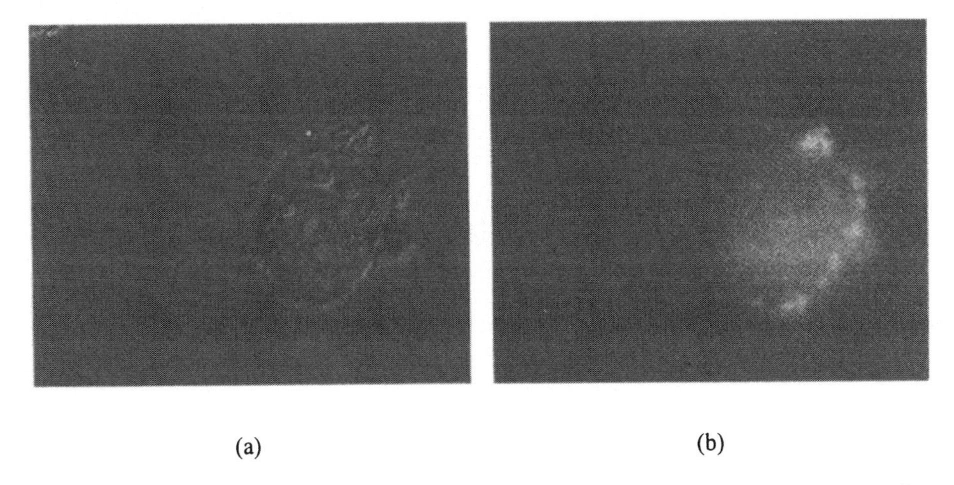

(a) (b)

Figures 8a. and 8b. (a) White light image of the fibroblast cells and (b) fluorescence microscope images of the CdS-functionalized fibroblast cells.

Clearly, the use of short peptides as linking agents is an effective tool for binding of fluorescent nanocrystals with nanoscale precision to the integrins on the fibroblast cells. It is important to realize that even a few nanometers can make an enormous difference in the nature of the interaction between the nanocrystals and the cellular integrins. Indeed, for intimate contact it is necessary to further reduce the nanocrystal-integrin separation.

5. CONCLUSION

This account has highlighted the recent progress in using semiconductor biotags based on their narrow, tunable and symmetric emission spectra as well as their temporal stability and resistance to photobleaching, especially as compared to fluorescent dyes. This progress has been possible as a result of key developments underlying the synthesis and functionalization of semiconductor nanocrystals. The advances in binding fluorescent semiconductor nanocrystals to biomolecules have facilitated the selective binding of these nanoscale fluorescent structures to specific subcellular structures. To go beyond using nanocrystals as biotags by integrating semiconductor nanocrystals directly with biological structures, it is necessary to further understand the physical properties of semiconductor nanocrystals in biological environments and in direct contact with biological structures. This review has highlighted several such interaction mechanisms, including the interaction of electrolytes with nanocrystals, the modification of the photoluminescence spectra of nanocrystals due to the environmentally-induced changes in the acoustic phonon spectra in nanocrystals, and the role of surface states on the observed intermittent blinking of quantum dots. To realize the possible uses of semiconductor nanocrystals as elements of coupled nanocrystal-biological-systems, it is necessary to study such interaction mechanisms in greater detail.

6. ACKNOWLEDGMENTS

The authors are appreciative of the encouragement and contributions of Aaron Johnson. The authors thank Prof. Philippe Guyot-Sionnest of the University of Chicago for identifying several key trends in quantum-dot research. In addition, the authors have benefited from discussions with Prof. Chad Mirkin of Northwestern University on the general topic of receptor-analyte interactions as well as with Prof. Raphael Tsu of the University of North Carolina of Charlotte on the topic of the Penn model as well as other related dielectric effects in quantum dots. Moreover, the authors are indebted to Drs. Todd Steiner and Dan Johnstone of the Air Force Office of Scientific Research as well as Dr. John Carrano of the Defense Advanced Research Projects Agency for their encouragement regarding our research on nanostructures. The authors are especially grateful to Dr. Dwight L. Woolard of the U.S. Army Research Office for his strong encouragement and support of research on the interaction of nanostructures with biological structures.

7. REFERENCES

1. Thomas M. Jovin, Quantum dots finally come of age, Nature Biotechnology 21, 32-33 (2003).
2. A. P. Alivisatos, Perspectives on the Physical Chemistry of Semiconductor Nanocrystals, J. Phys. Chem. 100, 13226-13239 (1996).
3. Erica Klarreich, Biologists Join the Dots, Nature 413, 450-452 (2001).
4. Jyoti K. Jaiswal, Hedi Mattoussi, J. Matthew Mauro, and Sanford M. Simon, Long term-multiple color imaging of live cells using quantum dot bioconjugates, Nature Biotechnology 21, 47-51 (2003); published online 2 December 2002.
5. Xingyong Wu, Hongjian Liu, Jianquan Liu, Kari N. Haley, Joseph A. Treadway, J. Peter Larson, Nianfeng Ge, Frank Peale, and Marcel P. Bruchez, Immunofluorescent labeling of cancer marker Her2 and other cellular targets with semiconductor quantum dots, Nature Biotechnology 21, 41-46 (2003).
6. Marcel Bruchez, Jr., Mario Moronne, Peter Gin, Shimon Weiss, Paul A. Alivisatos, Semiconductor Nanocrystals as Fluorescent Biological Labels, Science, 281, (1998) 2013-2016 (1998).
7. Warren C. W. Chan and Shuming Nie, Quantum Dot Bioconjugated for Ultrasensitive Nonisotropic Detection, Science 281, 2016-2018 (1998).
8. Mingyong Han, Xiaohu Gao, Jck Z. Su, and Shuming Nie, Quantum-dot-tagged microbeads for multiplexed optical coding of biomolecules, Nature Biotechnology 19, 631-635 (2001).
9. Michael A. Stroscio and Mitra Dutta, Advances in Quantum Dot Research and Technology: The Path to Applications in Biology, in Advanced Semiconductor Heterostructures, edited by Mitra Dutta and Michael A. Stroscio, (World Scientific Publ. Co., 2003), pages 101-118.
10. Gregoy P. Mitchell, Chad A. Mirkin, and Robert L. Letsinger, Programmed assembly of DNA functionalized quantum dots, Journal of the American Chemical Society 121 8122-8123 (1999).
11. Keith J. Watson, Jin Zhu, SonBinh T. Nguyen, and Chad A. Mirkin, Hybrid nanoparticles with block copolymer shell structures," Journal of the American Chemical Society 121 462-463 (1999).
12. YunWei Cao, Ronachao Jin, and Chad A. Mirkin, "DNA-modified core shell Ag/Au nanoparticles," Journal of the American Chemical Society 123 7961-7962 (2001).
13. Dimitri Alexson, Yang Li, Dinakar Ramadurai, Peng Shi, Leena George, Lenu George, Muzna Uddin, Preetha Thomas, Salvador Rufo, Mitra Dutta, and Michael A. Stroscio, Binding of semiconductor quantum dots to cellular integrins, accepted for publication in IEEE Transactions on Nanotechnology (2004).
14. Jessica O. Winter, Timothy Y.Liu, Brian A. Korgel, and Christine E. Schmidt, Recognition molecule directed interfacing between semiconductor quantum dots and nerve cells, Advanced Materials 13, 1673-1677 (2001).
15. Christine E. Schmidt, Venkatram R. Shastri, Joseph P. Vacanti, and Robert Langer, Stimultion of neurite outgrowth using an electrically conducting polymer, Proceedings of the National Academy of Sciences U.S.A. 94, 8948-8953 (1997).
16. H. M. Chen, X. F. Huang, L. Xu, J. Xu, K. J. Chen, and D. Feng, Self-assembly and photoluminescence of CdS-mercaptoacetic clusters with internal structures, Superlattices and Microstructures 27, 1-5 (2000).
17. A. Ekimov, Growth and optical properties of semiconductor nanocrystals, Journal of Luminescence 70, 1-20 (1996).
18. Yasushi Ohfuti, and Kikuo Cho, Size- and shape-dependent optical properties of assembled semiconductor spheres, Journal of Luminescence 70, 203-211 (1996).

19. O. I. Micic and A. J. Nozik, Synthesis and characterization of binary and ternary III-V quantum dots, Journal of Luminescence 70, 95-107 (1996).
20. T. Takagahara, Electron-phonon in semiconductor nanocrystals, Journal of Luminescence 70, 129-143 (1996).
21. L. Banyai and S. W. Koch, editors, Semiconductor Quantum Dots, Series of Atomic, Molecular, and Optical Physics, 2 (World Scientific, Singapore, New Jersey, London, Hong Kong, 1993).
22. D. Binberg, M. Grundmann, and N. N. Ledentson, Quantum Dot Heterostructures (John Wiley & Sons, Chichester, 1999).
23. U. Woggon, Optical Properties of Semiconductor Quantum Dots, Springer Tracts in Modern Physics 136 (Springer, Berlin, 1996).
24. V. Vasilevskiy, A. G. Rolo, and M. J. M. Gomes, One-phonon Raman scattering from arrays of semiconductor nanocrystals, Solid State Communications 104, 381-386 (1997).
25. R. Englman and R. Ruppin, Optical lattice vibrations in finite ionic crystals: I, Journal of Physics C (Proc. Phys. Soc.) 1, 614-629 (1968).
26. X. X. Liang and Jinsheng Yang, Effective-phonon approximation of polarons in ternary mixed crystals, Solid State Communications 100, 629-634 (1996).
27. Xin-Qi Li and Yasuhiko Arakawa, Optical linewidths in an individual quantum dots, Physical Review B60, 1915-1920 (1999).
28. L. Banyai, P. Gilloit, Y. Z. Hu, and S. W. Koch, Surface-polarization instabilities of electron-hole pairs in semiconductor quantum dots, Physical Review B45, 14136-14142 (1992).
29. Xin-Qi Li, Hajime Nakayama, and Yasuhiko Arakawa, Phonon bottleneck in quantum dots: Role of lifetime of the confined optical phonons, Physical Review B59, 5069-5073 (1999).
30. Igor Vurgaftman and Jasprit Singh, Effect of spectral broadening and electron-hole scattering on carrier relaxation in GaAs quantum dots, Applied Physics Letters 64, 232-234 (1994).
31. T. Takagahara, Theory of exciton dephasing in semiconductor quantum dots, Physical Review B60, 2638-2652 (1999).
32. L. P. Kouwenhoven, S. Jauhar, J. Orenstein, P. L. McEuen, Y. Nagamune, J. Motohisa, and H. Sakaki, Observation of phonon-assisted tunneling through a quantum dot, Physical Review Letters 73, 3443-3446 (1994).
33. Hideki Gotoh, Hiroaki Ando, and Toshihide Takagahara, Radiative recombination lifetime of excitons in thin quantum boxes, Journal of Applied Physics 81, 1785-1789 (1997).
34. Igor Vurgaftman, Yeeloy Lam, and Jasprit Singh, Carrier thermalization in sub-three-dimensional electronic systems: Fundamental limits on modulation bandwidth in semiconductor lasers, Physical Review B50, 14309-14326 (1994).
35. Xin-Qi Li and Yasuhiko Arakawa, Ultrafast energy relaxation in quantum dots through defect states: A lattice-relaxation approach, Physical Review B56, 10423-10427 (1997).
36. M. Grundmann, J. Christen, N. N. Ledentsov, J. Bohrer, D. Bimberg, S. S. Ruvimov, P. Werner, U. Richter, U. Gosele, J. Heydenreich, V. M. Ustinov, A. E. Zhukov, P. S. Kop'ev, and Zh. I. Alferov, Ultranarrow luminescence lines from single quantum dots, Physical Review Letters 74, 4043-4046 (1995).
37. Mohammad El-Said, Study of the energy level-crossings in GaAs/$Al_xGa_{1-x}As$ quantum dots, Solid State Communications 97, 971-971 (1996).
38. T. Inoshita and H. Sakaki, Electron-phonon intercation and the so-called phonon bottleneck effect in semiconductor quantum dots, Physics B 227, 373-377 (1996).
39. S. Komiyama, O. Astafiev, V. Antonov, T. Kutsuwa, and H. Hirai, A single-photon detector in the far-infrared range, Nature 403, 405-407 (2000).
40. Selvakumar V. Nair and T. Takagahara, Weakly correlated exciton pair states in large quantum dots, Physical Review B53, R10516-R10519 (1996).
41. M. Grundmann, O. Stier, and D. Bimberg, InAs/GaAs pyramidal quantum dots: Strain distribution, optical phonons, and electronic structure, Physical Review B52, 11969-11981 (1995).
42. J.-Y. Marzin, J.-M. Gerard, A. Izael, D. Barrier, and G. Bastard, Photoluminescence of single InAs quantum dots obtained by self-organized growth on GaAs, Physical Review Letters 73, 716-719 (1994).
43. P. Boucaud, K. S. Gill, J. B. Williams, M. S. Sherwin, W. V. Schoenfeld, and P. M. Petroff, Saturation of THz-frequency intraband absorption in InAs/GaAs quantum dot molecules, Applied

Physics Lettters 77, 510-512 (2000); P. Boucaud, J. B. Williams, K. S. Gill, M. S. Sherwin, W. V. Schoenfeld, and P. M. Petroff, Terahertz-frequency electronic coupling in vertically coupled quantum dots, Applied Physics Letters 77, 4356-4358 (2000).

44. S. Sauvage, P. Boucaud, F. H. Julien, J.-M. Gerard, and V. Thierry-Mieg, Intraband absorption in n-doped InAs/GaAs quantum dots, Applied Physics Letters 71, 2785-2787 (1997).
45. Hailong Zhou, A. V. Nurmikko, C.-C. Chu, J. Han, R. L. Gunshor, and T. Takagahara, Observation of antibound states for exciton pairs by four-wave-mixing experiments in a single ZnSe quantum well, Physical Review B58, R10131-R10134 (1998).
46. B. Legrand, S. Grandidier, J. P. Nyes, D. Stievenard, J. M. Gerard, and V. Thierry-Mieg, Scanning tunneling microscopy and scanning tunneling spectroscopy of self-assembled InAs quantum dots, Applied Physics Letters 73, 96-98 (1998).
47. S. Sauvage, P. Boucaud, F. Glotin, R. Prazeres, J.-M. Ortega, A. Lemaitre, J.-M. Gerard, V. Thierry-Mieg, Saturation of intraband and electron relaxation time in n-doped InAs/GaAs self-assembled quantum dots, Applied Physics Letters 73, 3818-3820 (1998).
48. B. Legrand, J. P. Nys, S. Grandidier, A. Lemaitre, J. M. Gerard, and V. Thierry-Mieg, Quantum box size effect on vertical self-alignment studied using cross-sectional scanning tunneling microscopy, Applied Physics Letters 74, 2608-2610 (1999).
49. S. Ruvimov, P. Werner, K. Scheerschmidt, U. Gosele, J. Heydenreich, U. Richter, N. N. Ledentsov, M. Grundmann, D. Bimberg, V. M. Ustinov, A. Yu Ergorov, P. S. Kop'ev, and Zh. I. Alferov, Structural charaterization of (In,Ga)As quantum dots in a GaAs matrix, Physical Review B51, 14766-14769 (1995).
50. T. Bruchez, P. Boucaud, S. Sauvage, F. Glotin, R. Prazeres, J.-M. Ortega, A. Lemaitre, and J.-M. Gerard, Midinfrared second-harmonic generation in p-type InAs/GaAs self-assembled quantum dots, Applied Physics Letters 75, 835-837 (1999).
51. S. Sauvage, P. Boucaud, T. Brunhes, V. Immer, E. Finkman, and J.-M. Gerard, Midinfrared absorption and photocurrent spectroscopy of InAs/GaAs self-assembled quantum dots, Applied Physics Letters 78, 2327-2329 (2001).
52. V. V. Mitin, V. I. Pipa, A. V. Sergeev, M. Dutta and M. Stroscio, High-gain quantum-dot infrared photodetector, Infrared Physics & Technology 42, 467-472 (2001).
53. V. Ryzhii, I. Khmyrova, V. Mitin, M. Stroscio, and M. Willander, On the detectivity of quantum-dot infrared photodetectors, Applied Physics Letters 78, 3523-3525 (2001).
54. J. Phillips, Pallab Bhattacharya, S. W. Kennerly, D. W. Beekman, and Mitra Dutta, Self-assembled InAs-GaAs quantum-dot intersubband detectors, IEEE Journal of Quantum Electronics 35, 936-943 (1999).
55. G. Yusa, and H. Sakaki, InAs quantum dot field effect transistors, Superlattices and Microstructures 25, 247-250 (1999).
56. R. L. Sellin, Ch. Ribbat, M. Grundmann, N. N. Ledentsov, and D. Bimberg, Close-to-ideal device characteristics of high-power InGaAs/GaAs quantum dot lasers, Applied Physics Letters 78, 1207-1209 (2001).
57. Ph. Lelong, K. Suzuki, G. Bastard, H. Sakaki, and Y. Arakawa, Enhancement of the Coulomb correlations in type-II quantum dots, Physica E 7, 393-397 (2000).
58. O. I. Micic, S. P., Ahrenkiel, D., Bertram, and A. J. Nozik, Synthesis, structure, and optical properties of colloidal GaN quantum dots, Applied Physics Letters 75, 478-480 (1999).
59. K. Tachibana, T. Someya, Y. Arakawa, R. Werner, and A. Forchel, Room-temperature lasing oscillation in an InGaN self-assembled quantum dot laser, Applied Physics Letters 75, 2605-2607 (1999).
60. Michael A. Stroscio and Mitra Dutta, Mitra, Damping of nonequilibrium acoustic phonon modes in a semiconductor quantum dot, Physical Review B60, 7722 (1999).
61. Michael A. Stroscio, K. W. Kim, SeGi Yu, and Arthur Ballato, Quantized acoustic phonons in quantum wires and quantum dots, Journal of Applied Physics 76, 4670-4975 (1994).
62. Todd D. Krauss, and Frank W. Wise, Coherent acoustic phonons in a semiconductor quantum dot, Physical Review Letters 79, 5102-5105 (1997).
63. Todd D. Krauss, Frank W. Wise, and David B. Tanner, Observation of coupled vibrational modes of a semiconductor nanocrystal, Physical Review Letters 76, 1376-1379 (1996).

64. P. Hoyer and R. Konenkamp, Photoconduction in porous TiO_2 sensitized PdS quantum dots, Applied Physics Letters 66, 349-351 (1995).
65. Brian L. Wehrenberg, Congjun Wang, and Phillip Guyot-Sionnest, Interband and intraband optical studies of PbSe colloidal quantum dots, J. Phys. Chem. B 106, 10634-10640 (2002).
66. G. Springholz, V. Holy, M. Pinczolits, and G. Bauer, Self-organized growth of three-dimensional quantum-dot crystals with fcc-like stacking and tunable lattice constant, Science 282, 734-737 (1998).
67. A. M. de Paula, L. C. Barbosa, C. H. B. Cruz, O. L. Alves, J. A. Sanjurjo, and C. L. Cesar, Quantum confinement effects on the optical phonons of CdTe quantum dots, Superlattices and Microstructures 23, 1104-1106 (1998).
68. C. R. M. de Oliveira, A. M. de Paula, F. O. Filho, Neto Plentz, J. A. Medeiros, L. C. Brabosa, O. L. Alves, E. A. Menezes, J. M. M. Rios, H. L. Fragnito, C. H. Brito Cruz, and C. L. Cesar, Probing of the quantum dot size distribution in CdTe-doped glasses by photoluminescence excitation spectroscopy, Applied Physics Letters 66, 439-441 (1995).
69. V. Esch, B. Fluegel, G. Khitrova, H. M. Gibbs, Xu Jiajin, K. Kang, S. W. Koch, L. C. Liu, S. H. Risbud, and N. Peyghambarian, State filling, Coulomb, and trapping effects in the optical nonlinearity of CdTe quantum dots in glass, Physical Review B42, 7450-7453 (1990).
70. O. I. Micic, S. P. Ahrenkiel, and Arthur J. Nozik, Synthesis of extremely small InP quantum dots and electronic coupling in their disordered solid films, Applied Physics Letters 78, 4022-4024 (2001).
71. Olga I. Micic, Julian Sprague, Zhenghao Lu, and Arthur J. Nozik, Highly efficient band-edge emission from InP quantum dots, Applied Physics Letters 68, 3150-3152 (1996).
72. U. Banin, G. Cerullo, A. A. Guzelian, C. J. Bardeen, A. P. Alivisatos, and C. V. Shank, Quantum confinement and ultrafast dephasing in InP nanocrystals, Physical Review Letters B55, 7059-7067 (1997).
73. H. Giessen, B. Fluegel, G. Mohs, N. Peyghambarian, J. R. Sprague, O. I. Mimic, and A. J. Nozik, Observation of the quantum confined ground state in InP quantum dots, Applied Physics Letters 68, 304-306 (1996).
74. J. J. Shiang, S.H. Rishbud, and A.P. Alivisatos, Resonance Raman studies of the ground and lowest electronic excited state in CdS nanocrystals, J. Chem. Phys. 98(11), 8432-8442 (1993).
75. one more cds
76. M. Danek, K. F. Jensen, C. B. Murray, and M. G. Bawendi, Electrospray organometallic chemical vapor deposition – A novel technique for preparation of II-VI quantum dot composites, Applied Physics Letters 65, 2795-2797 (1994).
77. M. Danek, Klavs F. Jensen, Chris B. Murray, and Moungi G. Bawendi, Synthesis of luminescent thin-film CdSe/ZnSe quantum dot composites using CdSe quantum dots passivated with an overlayer of ZnSe, Che. Mater. 8, 173-180 (1996).
78. E. Empedocles, D. J. Norris, and M. G. Bawendi, Photoluminescence spectroscopy of single CdSe nanocrystallite quantum dots, Physical Review Letters 77, 3873-3876 (1996).
79. J. R. Heine, J. Rodriguez-Viejo, M. G. Bawendi, and K. F. Jensen, K. F., Synthesis of CdSe quantum dot-ZnS matrix thin films via electrospray organometallic chemical vapor deposition, Journal of Crystal Growth 195, 564-568 (1998).
80. Hedi Mattoussi, J. Matthew Mauro, Ellen R. Goldman, George P. Anderson, Vikrim C. Sundar, Frederic V. Mikulec, and Moungi G. Bawendi, Self-assembly of CdSe-ZnS quantum dot bioconjugates using an engineered recombinant protein, Journal of the Americam Chemical Society 122, 12142-12150 (2000).
81. Xiaogang Peng, Liberato Manna, Weidong Yang, Juanita Wickham, Erik Scher, E. Kadavanich, and A. P. Alivisatos, Shape control of CdSe nanocrystals, Nature 404, 59-61 (2000).
82. Chia-Chun Chen, A. B. Herhold, C. S. Johnson, and A. P. Alivisatos, Size dependence of structural metastability in semiconductor nanocrystals, Science 276, 398-401 (1997).
83. D. E. Fogg, L. H. Radziloski, B. O. Dabbousi, R. R. Schrock, E. L. Thomas, and M. G. Bawendi, Fabrication of quantum dot-polymer composites: Semiconductor nanoclusters in dual-function polymer matrices with electron-transporting and cluster-passivating properties, Macromolecules 30, 8433-8439 (1997).

84. Margaret Hines and Philippe Guyot-Sionnest, Synthesis and characterization of strongly luminescing ZnS-capped CdSe nanocrystals, Journal of Physical Chemistry 100, 468-471 (1996).
85. Phedon Palinginis and Hailin Wang, High-resolution spectral hole burning in CdSe/ZnS core/shell nanocrystals, Applied Physics Letters 78, 1541-1543 (2001).
86. C. E. Finlayson, D. S. Ginger, and Neil C. Greenham, Optical microcavities using highly luminescent films of semiconductor nanocrystals, Applied Physics Letters 77, 2500-2502 (2000).
87. Ehud Poles, Donald C. Selmarten, Olga I. Micic, and Arthur J. Nozik, Anti-Stokes photoluminescence in colloidal semiconductor quantum dots, Applied Physics Letters 75, 971-973 (1999).
88. Philippe Guyot-Sionnest, and Margaret A. Hines, Intraband transition in semiconductor nanocrystals," Applied Physics Letters 72, 686-688 (1998).
89. Mark E. Schmidt, Sean A. Blanton, Margaret A. Hines, and Philippe Guyot-Sionnest, Polar CdSe nanocrystals: Implications for electronic structure, Journal of Chemical Physics 106, 5254-5259 (1997).
90. B. O. Dabbousi, M. G. Bawendi, O. Onitsuka, and M. F. Rubner, Electroluminescence from CdSe quantum-dot/polymer composites, Applied Physics Letters 66, 1316-1318 (1995).
91. D. J. Norris, A. Sacra, C. B. Murray, and M. G. Bawendi, Measurement of the size dependent hole spectrum in CdSe quantum dots, Physical Review Letters 72, 2612-2615 (1994).
92. U. Woggon, S. Gaponenko, W. Langbein, A. Uhrig, and C. Klingshirn, Homogeneous linewidth of confined electron-hole-pair states in II-VI quantum dots, Physical Review B47, 3684-3689 (1993).
93. Shinrtaro Nomura, and Takayoshi Kabayashi, Exciton-LO-phonon couplings in spherical microcrystallites, Physical Review B45, 1305-1316 (1992).
94. M. C. Klein, F. Hache, D. Ricard, and C. Flytzanis, Size dependence of electron-phonon coupling in semiconductor nanospheres: The case of CdSe, Physical Review B42, (1990) 11123-11132 (1990).
95. Gaetano Scamarcio, Mario Lugara, and Daniela Manno, Size-dependent lattice constant in $CdS_{1-x}Se_x$ nanocrystals embedded in glass observed by Raman scattering, Physical Review B45, 13792-13795 (1992).
96. T. Takagahara, Electron-phonon interactions and exciton dephasing in semiconductor nanocrystals, Physical Review Letters 71, 3577-3580 (1993).
97. R. W. Schoenlein, D. M. Mittleman, J. J. Shiang, A. P. Alivisatos, and C. V. Shank, Investigation of femtosecond electronic dephasing in CdSe nanocrystals using quantum-beat-suppressed photon echoes, Physical Review Letters 70, 1014-1017 (1993).
98. D. M. Mittleman, R. W. Schoenlein, J. J. Shiang, V. L. Colvin, A. P. Alivisatos, and C. V. Shank, Quantum size dependence of femtosecond electronic dephasing and vibrational dynamics in CdSe nanocrystals, Physical Review B49, 14435-14447 (1994).
99. Sean A. Blanton, Margaret A. Hines, and Philippe Guyot-Sionnest, Photoluminescence wandering in single CdSe nanocrystals, Applied Physics Letters 69, 3905-3907 (1996).
100. J. Rodriguez-Viejo, K. F. Jensen, H. Mattoussi, J. Michel, B. O. Dabbousi, and M. G. Bawendi, Cathodoluminescence and photoluminescence of highly luminescent CdSe/ZnS quantum dot composites, Applied Physics Letters 70, 2132-2134 (1997).
101. Boaz Alperson, Israel Rubinstein, and Gary Hodges, Energy level tunneling spectroscopy and single electron charging in individual CdSe quantum dots, Applied Physics Letters 75, 1751-1753 (1999).
102. V. I. Klimov, Ch. J. Schwarz, D. W. McBranch, C. A. Leatherdale, and M. G. Bawendi, Ultrafast dynamics of inter- and intraband transitions in semiconductors nanocrystals: Implications for quantum-dot lasers, Physical Review B60, R2177-R2180 (1999).
103. Moonsub Shim and Philippe Guyot-Sionnest, n-type colloidal semiconductor nanocrystals, Nature 407, 981-983 (2000).
104. Congjun Wang, Moonsub Shim, and Philippe Guyot-Sionnest, Electrochromic nanocrystal quantum dots, Science 291 2390-2392 (2001).
105. M. G. Bawendi, W. L. Wilson, L. Rothberg, P. J. Carroll, T. M. Jedju, M. L. Steigerwald, and L. E. Brus, Electronic structure and photoexcited-carrier dynamics in nanometer-size CdSe clusters, Physical Review Letters 65, 1623-1626 (1990).

106. Sean A. Blanton, Robert L. Leheny, Margaret A. Hines, and Philippe Guyot-Sionnest, Dielectric dispersion measurements of CdSe nanocrystal colloids: Observation of a permanent dipole moment, Physical Review Letters 79, 865-868 (1997).
107. David L. Klein, Richard Roth, Andrew K. L. Lim, A. Paul Alivisatos,, and Paul L. McEuen, A single-electron transistor made from a cadmium selenide nanocrystal," Nature 389 (1997) 699-701.
108. D. S. Ginger, A. S. Dhoot, C. E. Finlayson, and N. C. Greenham, Long-lived quantum-confined infrared transitions in CdSe nanocrystals, Applied Physics Letters 77, 2816-2818 (2000).
109. Moonsub Shim and Philippe Guyot-Sionnest, n-type colloidal semiconductor nanocrystals, Nature 407, 981-983 (2000).
110. Sean A. Blanton, Robert L. Leheny, Margaret A. Hines, and Philippe Guyot-Sionnest, Dielectric dispersion measurements of CdSe nanocrystal colloids: Observation of a permanent dipole moment, Physical Review Letters 79, 865-868 (1997).
111. Pallab Bhattacharya, Quantum Dot Semiconductor Lasers, in Advances in Semiconductor Lasers and Applications to Optoelectronics, Selected Topics in Electronics and Systems, 16, eds., Mitra Dutta and Michael A. Stroscio (World Scientific, Singapore, New Jersey, London, Hong Kong, 2000), pp. 235-261.
112. A. E. Zhukov, V. M. Ustinov, and Z. I. Alferov, Device characteristics of low-threshold quantum-dot lasers, in: Advances in Semiconductor Lasers and Applications to Optoelectronics, Selected Topics in Electronics and Systems, 16, eds. Dutta, Mitra and Stroscio, Michael A. (World Scientific, Singapore, New Jersey, London, Hong Kong, 2000), pp. 419-431.
113. M. Dutta and M. A. Stroscio, Advanced semiconductor lasers: Phonon engineering and phonon interactions, in: Advances in Semiconductor Lasers and Applications to Optoelectronics, SelectedTopics in Electronics and Systems, 16, eds. Dutta, Mitra and Stroscio, Michael A. (World Scientific, Singapore, New Jersey, London, Hong Kong, 2000), pp. 419-431.
114. A. D. Andreev and E. P. O'Reilly, Theory of the electric structure of GaN/AlN hexagonal quantum dots, Phys. Rev. B62, 15851-15870 (2001).
115. J. Jerphagnon, Imvariants of the third-rank Cartesian-tensor: Optical nonlinear susceptibilities, Phys. Rev. B2, 1091-1098 (1970).
116. Christina J. Zhang, Howard W. H. Lee, Ian M. Kennedy, and Subhash H. Risbud, Observation of quantum confined excited states in GaN nanocrystals, Appl. Phys. Lett. 72, 3035-3037 (1998).
117. Sandra J. Rosenthal, Ian Tomlinson, Erika M. Adkins, Sally Schroeter, Scott Adams, Laura Swafford, James McBride, Yongqiang Wang, Louis J. DeFelic, and Randy D. Blakey, Targeting cell surface receptors with ligand-conjugated nanocrystals, Journal of the American Chemical Society 124 4586-4594 (2000).
118. Maria E. Akerman, Warren C.W. Chan, Pirjo Laakkonen, Sangeeta N. Bhatia, and Erkki Ruoslahti, Nanocrystal targeting in vivo, Proceedings of the National Academy of Sciences 99 (20), 12617-12621 (2002).
119. Hedi Mattoussi, J. Matthew Mauro, Ellen R. Goldman, George P. Anderson, Vikram C. Sundar, Frederic V. Mikulec, and Moungi G. Bawendi, Self-assembly of CdSe-ZnS quantum dot bioconjugates using an engineered recombinant protein, J. Am. Chem. Soc. 122 12142-12150 (2000).
120. E. R. Goldman, G. P. Anderson, P. T. Tran, H. Mattousi, P. T. Charles, and J. M. Mauro, Conjugation of luminescent quantum dots with antibodies using an engineered adaptor protein to provide new reagents for fluoroimmunoassays, Anal. Chem. 74 841-847 (2002).
121. W.C. Chan and S. Nie, Quantum dot bioconjugates for ultrasensitive nonisotopic detection, Science 281, 2016-2018 (1998).
122. Maxime Dahan, Sabine Levi, Camilla Luccardini, Philippe Rostaing, Beatrice Riveau, and Antoine Triller, Diffusion dynamics of glycine receptors revealed by single quantum dot tracking, Science 302, 442-445 (2003).
123. Michael A. Stroscio and Mitra Dutta, Phonons in Nanostructures (Cambridge University Press, Cambridge, 2001).
124. R. M. de la Cruz, S. W. Teitsworth, and M. A. Stroscio, Phonon bottleneck effects for confined longitudinal optical phonons in quantum boxes, Superlattices and Microstructures, 13, (1993) 481-486.

125. R. M. de la Cruz, S. W. Teitsworth, and M. A. Stroscio, Interface Phonons in spherical GaAs/Al(x)Ga(1-x)As quantum dots, Physical Review B52 1489-1492 (1995).
126. V. Mitin, V. Kochelap, and Michael Stroscio, Quantum Heterostructures for Microelectronics and Optoelectronics (Cambridge University Press, Cambridge, 1999).
127. M. P. Chamberlain, C. Trallero-Giner, and M. Cardona, Theory of one-phonon Raman scattering in semiconductor microcrystallites, Physical Review B51, 1680-1683 (1995).
128. J. C. Marini, B. Stebe, and E. Kartheuser, Exciton-phonon interaction in CdSe and CuCl polar semiconductor nanospheres, Physical Review B50, 14302-14311 (1994).
129. E. Roca, C. Trallero-Giner, and M. Cardona, Polar optical vibrational modes in quantum dots, Physical Review B49, 13704-13711 (1994).
130. P. A. Knipp and T. L. Reinecke, Classical interface modes in quantum dots, Physical Review B46, 10310-10319 (1992).
131. Dmitri Romanov, Vladimir Mitin, and Michael Stroscio, Polar surface vibration strips on GaN/AlN quantum dots and their interaction with confined electrons, Physica E 12, 491-494 (2002).
132. Augusto M. Alcalde and Gerald Weber, Scattering rates due to electron-phonon interaction in $CdS_{1-x}Se_x$ quantum dots, Semiconductor Science and Technology 15, 1082-1086 (2000).
133. C. A. Leatherdale, and M. G. Bawendi, Observation of solvatochromism in CdSe colloidal quantum dots, Physical Review B63, 165315-1-6 (2001).
134. C. A. Leatherdale, C. R., Kagan, N. Y. Morgan, S. A. Empedocles, M. A. Kastner, and M. G. Bawendi, Photoconductivity in CdSe quantum dot solids, Physical Review B62, 2669-2679 (2000).
135. Raphael Tsu, and Davorin Babic, Doping of a quantum dot, Applied Physics Letters 64, 1806-1808 (1994).
136. D. B. Tran Thoai, Y. Z. Hu, S. W. Koch, Influence of the confinement potential on the electron-hole-pair states in semiconductor microcrystallites, Physical Review B42, 11261-11266 (1990).
137. Raphael Tsu, Davorin Babic, and Liderio Ioriatti, Simple model for the dielectric constant of nanoscale silicon particle, Journal of Applied Physics 82, 1327-1329 (1997).
138. Lin-Wang Wang, and Alex Zunger, Pseudopotential calculations of nanoscale CdSe quantum dots, Physical Review B53, 9579-9582 (1996).
139. David R. Penn, Wave-mumber-dependent dielectric function of semiconductors, Physical Review 128, 2093-2097 (1992).
140. Davoran Babic, Raphael Tsu, and Richard F. Green, Ground-state energies of one- and two-electron silicon dots in an amorphous silicon dioxide matrix, Physical Review B45, 14150-14155 (1992).
141. A. Tamura, K. Higeta, and T. Ichinokawa, Lattice vibrations and specific heat of a small particle, Journal of Physics C: Solid State Physics 15, 4975-4991 (1982).
142. Shaklee, Fred H. Pollak, and Manuel Cardona, Electroreflectance at a semiconductor-electrolyte interface, Phys. Rev. Lett. 15(23) 883-885 (1965).
143. M. Kuno, D. P. Fromm, H. F. Hamann, A. Gallagher, and D. J. Nesbitt, Nonexponential "blinking" kinetics of single CdSe quantum dots: a universal power law behavior, J. Chem. Phys. 112(7), 3117-3120 (2000).
144. Mitsura Sugisaki, Hong-Wen Ren, Kenichi Nishi, and Yasuaki Masumoto, Fluorescence intermittency in self-assembled InP quantum dots, Phys. Rev. Lett., 856(21), 4883-4886 (2001).
145. YounJoon Jung, Eli Barkai, and Robert J. Silbey, Lineshape theory and phonon counting statistics for blinking quantum dots: a Levy walk process, Chemical Physics 284, 181-194 (2002).
146. Harvey Scher and Elliott W. Montrol, Anomalous transit-time dispersion in amorphous solids, Physical Review B12(6), 2455-2477 (1975).
147. B. Schreder, C. Dem, M. Schmitt, A. Materny, W. Kiefer, U. Winkler, and E. Umback, Raman spectroscopy of II-VI semoconductor nanostructures: CdS quantum dots, Journal of Raman Spectroscopy 34, 100-103 (2003).
148. Fabio Bernardini, Vincenzo Fiorentini, and David Vanderbilt, Spontaneous polarization and piezoelectric constants of III-V nitrides, Phys. Review B56, R10024-R10127 (1997).
149. A. L. Hodgkin, The Conduction of the Nervous Impulse (Liverpool University Press, Liverpool, 1964).
150. Jonathan Howard, Mechanics of Motor Proteins and the Cytoskeleton (Sinauer Associates, Inc. – Publishers, Sunderland, Massachusetts, 2001).

151. Erkki Ruoslahti, RGD and other recognition sequences for integrins, Annual Reviews of Cell & Developmental Biology 12(1), 697-715 (1996).
152. Gerald Karp, Cell and Molecular Biology, 3rd Ed. (Wiley, 2002).
153. Robert P. Lanza, Robert Langer, and Joseph Vacanti, editors, Principles of Tissue Engineering (Academic Press, 2000).
154. Bruce Alberts, Alexander Johnson, Julian Lewis, Martin Raff, Keith Roberts, and Peter Walters, Molecular Biology of the Cell, 4th Ed. (Garland Science, Taylor & Francis Group, 2001).
155. H. K. Kleinman, B. S. Weeks, F. B. Cannon, T. M. Sweeney, G. C. Sephel, B. Clement, M. Zain, M. O. J. Olson, M. Juncker, B. A. Burrous, Identification of a 110-kDa non-integrin cell surface laminin-binding protein which recognizes an A chain neurite-promoting peptide, Arch. Biochem. Biophys. 2909, 320-325 (1991).
156. Sharon K. Powell and Hynda Kleinman, Neuronal laminins and their cellular receptors, Int. J. Biochem. Cell. Biol. 29, 401-414 (1997).
157. Sharon K. Powell, Javashree Rao, Eva Roque, Motoyoshi Nomizu, Yuichiro Kuratomi, Yoshihiko Yamada, and Hynda K. Kleinman, Neural cell response to multiple novel sites on laminin-1, Journal of Neuroscience Research 61, 302-312 (2000).
158. Dimitri Alexson, Yang Li, Dinakar Ramadurai, Peng Shi, Leena George, Lenu George, Muzna Uddin, Preetha Thomas, Salvador Rufo, Mitra Dutta, and Michael A. Stroscio, Binding of semiconductor quantum dots to cellular integrins, IEEE Transactions on Nanotechnology, in press (2004).
159. Masao Iwamatsu, Makoto Fujiwara, Naoisa Happo, and Kenju Horii, Effects of dielectric discontinuity on the ground-state energy of charged Si dots covered with a SiO_2 layer, Journal of Physics: Condensed Matter 9, 9881-9892 (1997); Masao Iwamats and Kenju Horii, Dielectric confinement effects on the impurity and exciton binding energies of silicon dots covered with a silicon dioxide layer, Japanese Journal of Applied Physics 36, 6416-6423 (1997).
160. Anand Venkatesan, MS Thesis, Univesity of Illinois at Chicago (2003).
161. Dinakar Ramadurai, Babak Kohanpour, Dimitri Alexson, Peng Shi, Akil Sethuraman, Yang Li, Vikas Saini, Mitra Dutta, and Michael A. Stroscio, Tunable optical properties of colloidal quantum dots in electrolytic environments, IEE Proceedings in Nanobiotechnology, in press (2004).
162. P. E. Lippens and M. Lannoo, Calculation of the band gap for small CdS and ZnS crystallites, Phys. Rev. B39, 10935-10942 (1989).
163. A. S. Davydov, Quantum Mechanics (NEO Press, Ann Arbor, 1966).
164. Salvador Rufo, Mitra Dutta, and Michael A. Stroscio, "Acoustic modes in free and embedded quantum dots," Journal of Applied Physics 93, 2900-2905 (2003).

BIOMEDICAL APPLICATIONS OF SEMICONDUCTOR QUANTUM DOTS

Anupam Singhal, Hans C. Fischer, Johnson Wong, Warren C. W. Chan*

1. INTRODUCTION

In recent decades, the exquisite sensitivity and versatility of optical technologies have led to numerous breakthroughs in biological research, including real-time imaging of live cells,[1, 2] gene expression profiling,[3] cell sorting,[4] and clinical diagnostics.[5, 6] A key component in optical detection schemes is the probe design. These probes are constructed from organic fluorophores, such as fluorescein and tetramethylrhodamine (TMR), and recognition molecules. The optical emission of fluorophores is used to visualize the activities of biomolecules, while the recognition molecules direct the fluorophores to specific cells, tissues, or organs. Although optical probes are widely used, most organic fluorophores exhibit unfavourable properties that have hampered their applications in single-protein tracking in living cells, molecular pathology, and other research areas.[7] These properties include photobleaching, sensitivity to environmental conditions, and inability to excite multiple fluorophores using a single wavelength. A new generation of probes has emerged in the last five years that overcomes many of the limitations associated with organic fluorophores. These probes employ fluorophores that are sub-100 nm in size and composed of inorganic atoms. Unlike organic-only fluorophores, the optical and electronic properties of inorganic fluorophores can be tuned during the synthesis process by changing their size, shape, or composition. In this chapter, we will describe the use of one type of inorganic fluorophore, semiconductor nanocrystals, for the development of "custom-designed" probes for biomedical detection.

Semiconductor nanocrystals, also known as "quantum dots" (qdots), are typically composed of atoms from groups II-VI (CdSe, CdS, ZnSe) and III-V (InP and InAs), and are defined as particles with physical dimensions smaller than the Bohr exciton radius. The Bohr exciton radius of prototypical CdSe qdots, as illustrated in Fig. 1, is ~10 nm. The unique optical and electronic properties of qdots have spurred a great deal of research into their potential applications in the design of novel biological probes,[8, 9] light emitting diodes,[10, 11] photovoltaic cells,[12, 13] among other devices.

---

* A.S., H.C.F., J. W., W. C. W. C., Institute of Biomaterials and Biomedical Engineering, University of Toronto, Toronto, Ontario, Canada M5S 3G9. H.C.F. and W.C.W.C. are also affiliated with Department of Materials Science & Engineering, University of Toronto.

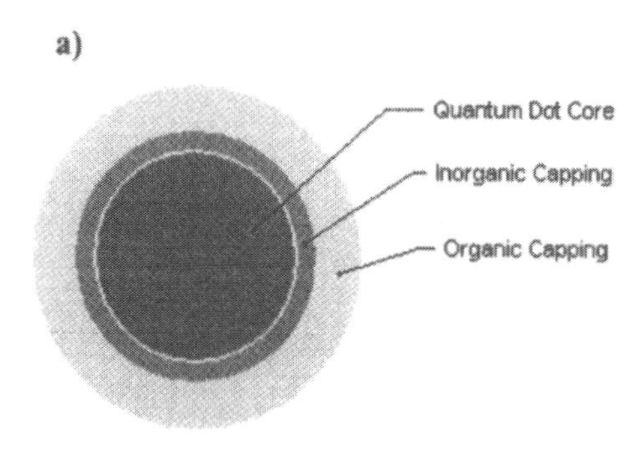

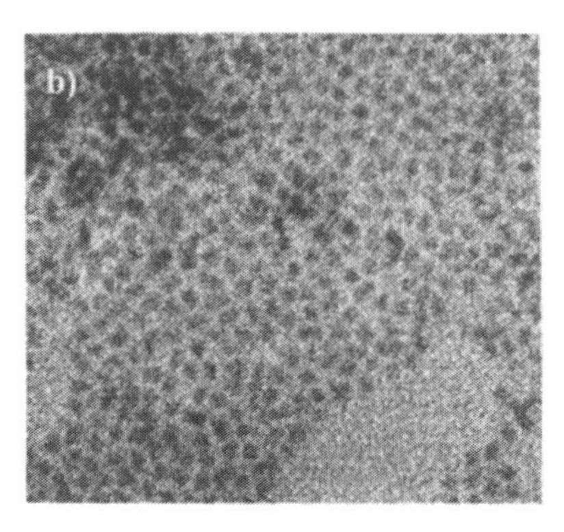

Figure 1. a) Schematic representation of quantum dots commonly used for biological labelling. b) Transmission electron micrograph (TEM) of monodisperse CdSe quantum dots (magnification = 200,000X).[14]

In this chapter, we focus on the biological applications of semiconductor qdots, beginning with a brief description of their unique optical and electronic properties. We then highlight various methods of synthesizing and characterizing qdots, as well as successful biological applications of these fluorophores. Finally, we discuss some of the prospects and challenges associated with future biological applications of qdots.

2. OPTICAL AND ELECTRONIC PROPERTIES OF SEMICONDUCTOR QUANTUM DOTS

Quantum dots (Qdots) exhibit unique optical and electronic properties that are only observed in an intermediate size regime between the size of discrete atoms and that of bulk solids. In this nanometre-size regime, charge carriers (i.e. electrons and holes) are spatially confined within the dimensions of qdots, a phenomenon known as quantum confinement. Due to this effect, the optical properties of qdots are heavily dependent upon their size, shape, composition, and surface interactions with their local environment. The use of excitation energy (e.g. via incident UV light) exceeding the bandgap energy of the qdots leads to the promotion of electrons from the valence band (ground state) to the conduction band (excited state), creating mobile electrons and holes. The light-induced excitation (or mobility) of electrons leads to fluorescence light emission in a process called radiative recombination. In this process, the electrons and holes interact to form an electron-hole pair called an exciton. The excited-state lifetimes of nanocrystals are multi-exponential with lifetimes of 5 ns, 20-30 ns, and 80-200 ns, with the 20-30 ns dominating. These processes can be measured using UV-Vis spectrophotometry or spectrofluorimetry.

Typical UV-Vis absorbance measurements of qdots produce broad, continuous spectra, which are dependent on the physical dimensions and composition of the particles as illustrated in Fig. 2a. One characteristic feature of the qdot absorbance curves is an observable peak, called the "quantum confinement" peak, which represents the lowest bandgap energy transition. A second characteristic feature of the qdot absorbance curves is the increasing absorbance at wavelengths shorter than the quantum confinement wavelength. This property is extremely advantageous for biomedical applications since qdots of all types and peak emission wavelengths can be excited using a single wavelength.

The fluorescence emission spectra of qdots are narrow and symmetric as exemplified by Fig. 2b. As with the qdot absorbance spectra, the fluorescence peak emission of qdots is heavily dependent on qdot composition and physical dimensions, with an observable red-shift in the peak fluorescence emission of larger qdots. The excitonic fluorescence emission for a bulk measurement of capped qdots (e.g., ZnS-capped CdSe) typically exhibits a full-width at half-maximum (FWHM) of ~30 nm and quantum yield of 20-50%. Conversely, measurements of the fluorescence spectra of single qdots have demonstrated a FWHM of 13 nm, ~ 2.5 times narrower than the typical bulk measurement. Since the fluorescence emission peak is size-tuneable, the broadness of bulk measurements can be attributed to size distributions of the qdots within the bulk solution. In comparison to qdots, many organic dye molecules (e.g. Rhodamine 6G) have broad, asymmetric emission spectra with FWHM > 45 nm as depicted in Fig. 3.

The fluorescence quality of qdots is often quantified in terms of a fluorescence quantum yield - the ratio between the number of fluorescence photons emitted when mobile electrons recombine with holes to the number of photons absorbed upon excitation. Defect structures both in the internal structure and on the surface of qdots can produce competing energy states that trap the excited electrons and holes, resulting in lower quantum yields. In a classic study by Alivisatos and coworkers,[15] the oxidative decomposition of CdSe qdots was shown to produce a broad-fluorescence peak that was red-shifted from the excitonic fluorescence peak. In their experiment, the oxidation of qdots produced an excess of unbonded-atoms (dangling bonds) on the qdot surface; this created low-energy bands to trap the mobile electrons and holes. In correspondence to the increase in the intensity of the red-shifted defect peak, the excitonic fluorescence intensity decreased.

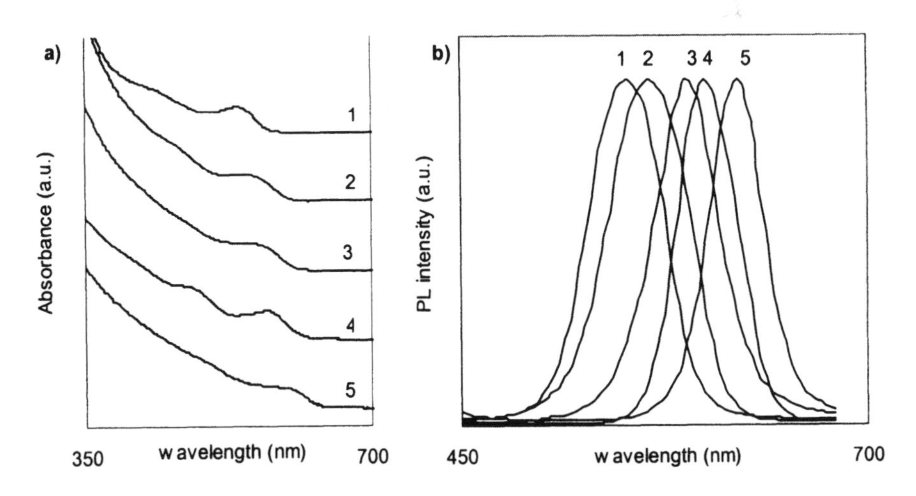

Figure 2. a) Absorbance spectra and b) Fluorescence Spectra of CdSe/ZnS quantum dots of five different sizes/emission colours (1 green, 2 yellow, 3 orange, 4 orange-red, 5 red-emitting).[14]

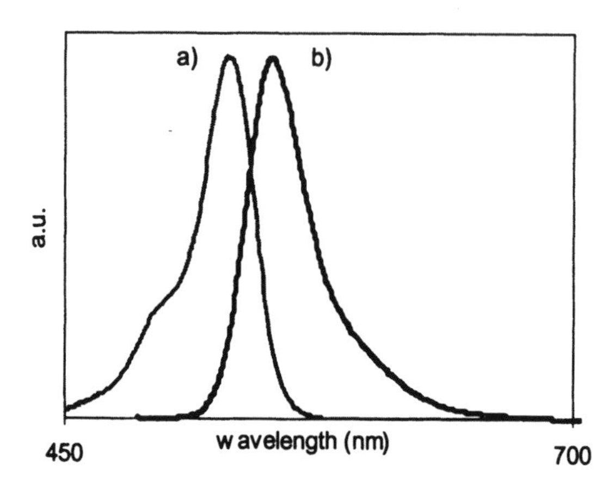

Figure 3. a) Absorbance and b) fluorescence spectra of the organic dye molecule Rhodamine 6G.

Approaches to improving the quantum yield of qdots have focused on removal of internal and surface defect structures. Internal defect structures can be removed by altering the solvent reaction conditions. For instance, while the presence of molecules such as H_2O and O_2 produces qdots with low quantum yields, the use of organic solvents and a controlled reaction atmosphere in a Schlenk line or glovebox have improved the quality of synthesized qdots (quantum yields > 5-10 %). Apart from internal defect structures, a major challenge to improving the quantum yield of qdots is to remove surface defect sites. Qdots have a large surface area-to-volume ratio and therefore, a large population of atoms on their surface. Removal of dangling bonds with organic stabilizing molecules and/or a second semiconductor layer has shown to dramatically improve the overall quantum yield of the qdots.[16] Careful selection of surface coating has produced qdots with quantum yields as high as 85%. In Section 3 - Synthesis and Characterization of Quantum Dots, strategies for coating qdots will be discussed in greater detail.

Two strategies for tuning the fluorescence emission, or bandgap energy, of quantum dots are the alteration of the size and composition of qdots. The bandgap energy is directly related to the composition of the qdot; for instance, CdS qdots exhibit ultraviolet (UV) fluorescence, while InP qdots exhibit near-IR fluorescence. Alloying qdots can also alter their bandgap energy. Recently, Bailey and coworkers demonstrated the shift in fluorescence emission of CdSe qdots to the near-IR by doping with Te.[17] Apart from altering qdot composition, the fluorescence emission of qdots of a single composition (e.g., CdSe) can be tuned by changing their size. As described by the quantum mechanical "particle-in-a-box" model, decreasing the size of qdots results in a corresponding increase in semiconductor bandgap energy (or, equivalently, shorter emission wavelengths). For example, the fluorescence emission peaks of cadmium selenide (CdSe) semiconductor qdots can be tuned across the entire visible spectrum from blue to red by increasing the diameter of the qdots from roughly 2 nm to 6 nm. In the "particle-in-a-box" model, the potential energy is infinite outside of the box and hence the particle is confined to the dimensions of the box. This particle contains discrete energy levels and wavefunctions that correspond to the dimensions of the box. This model coincides with the structure of qdots, in which mobile carriers (the particle) are confined within the dimensions of the qdot (the box) with discrete emission wavelengths (wavefunctions) and bandgap energy levels. As the physical dimensions of the box become smaller, the bandgap energy increases.

The fluorescence emission of qdots is extremely stable upon photoexcitation. In comparative measurements of qdots with small organic fluorophores, qdots have been shown to be ~ 100 times more stable against photobleaching. Qdots have also been shown to be more photostable than fluorescent proteins (e.g., phycoerytherin), although this result has not been quantified.[18] The photobleaching of qdots is believed to arise from a slow process of photo-induced chemical decomposition. This hypothesis is supported by our observation of a shift in emission-colour from red to blue when qdot-aggregates are spread on a glass slide and monitored under an epifluorescence microscope with high-power UV-excitation. Henglein and coworkers speculated that CdS decomposition is initiated by the formation of S or SH radicals upon optical excitation.[19] These radicals can react with O_2 from the air to form an SO_2 complex, resulting in slow particle degradation. Capping the surface of qdots with a thick second semiconductor layer has produced qdots with excellent photostability.

Due to the bright luminescence and high photostability of qdots, single qdots can be easily visualized and imaged under a conventional epifluorescence microscope. In addition, qdots are brighter than most small organic dye molecules due to their large molar extinction coefficients. Bawendi and coworkers[20] estimated that the molar extinction coefficients of CdSe qdots are about 10^5 to 10^6 M^{-1} cm^{-1}, depending on the particle size and the excitation wavelength. These values are 10-100 times larger than those of organic dyes, but are similar to the absorption cross sections of phycoerytherin, a multi-chromophore fluorescent protein. It has been estimated that a single qdot is approximately equivalent in fluorescence intensity to 10 to 20 small organic dye molecules.[21]

In this section, we have briefly described some of the properties of qdots that make them appealing for biological applications. For a more in-depth look at the physical chemistry of qdots, the interested reader can refer to a number of excellent reviews by Alivisatos[22], and Murray et al. [23]

3. SYNTHESIS AND CHARACTERIZATION OF QUANTUM DOTS

The most commonly synthesized semiconductor nanocrystals are composed of atoms from groups II-VI (e.g. CdSe, CdTe, CdS, and ZnSe) and groups III-V (e.g. InP and InAs) of the periodic table.[24-29] In particular, rapid advancements in synthesis and characterization techniques have led to the development of highly luminescent and monodisperse CdSe qdots.[16, 30-32] In the following sections, we will discuss a number of different approaches to the synthesis and characterization of high-quality qdots.

3.1. Synthesis Techniques

Typical techniques for the synthesis of semiconductor qdots involve the growth of nanocrystals using molecular precursors in either aqueous or organic solutions. In one approach to qdot synthesis in aqueous media, solutions of cadmium and sulphur precursors are injected into hot aqueous solutions containing stabilizing agents or in inverse micelles.[33,34] The size of the resulting nanocrystals can be controlled through the use of different solvent additives, or varying the solvent pH and temperature. Another approach to the synthesis of qdots in aqueous media employs yeast cells, such as *Candida glabrata* or *Schizosacharomyces pombe*.[35, 36] In the presence of cadmium or zinc ions, these yeast cells express proteins with thiol and carboxylic acid residues that bind to the metal ions and induce nucleation of nanocrystals. Despite the success and simplicity of both these aqueous methods, the resulting nanocrystals have low quantum yields (QY < 10%) and large size distributions (relative standard deviation RSD >15%), resulting in broad emission spectra (~50 nm full-width at half maximum, FWHM).

Improved synthesis schemes in organic media have led to the development of qdots that are highly luminescent (quantum yield > 50%) and monodisperse (relative size distributions < 5%).[16, 30-32] In one approach, qdots are generated by the pyrolysis of organometallic and chalcogenide precursors injected into a hot coordinating solvent. For instance, CdSe qdots are synthesized through the dissolution of dimethyl cadmium and selenium shot in either tri-n-butylphosphine (TBP) or tri-n-octylphosphine (TOP) and subsequent injection into a solution of tri-n-octylphospine oxide (TOPO) at 340-360°C. Rapid nucleation and growth of the nanocrystals are observed through changes in colour of the reaction mixture.

As described by La Mer and Dinegar,[37] nucleation occurs until the temperature and precursor concentrations drop below the critical "nucleation threshold". Subsequent re-heating of the nanocrystals to intermediate temperatures (250-300°C) results in slow growth of the nanocrystals. Growth of nanocrystals proceeds until the available precursor material is consumed, after which smaller nanocrystals begin to dissolve in order to supply materials for the growth of larger nanocrystals. This diffusion process, known as Ostwald ripening, occurs due to the high surface free energy of smaller nanocrystals, which make them more prone to dissolution than larger nanocrystals. This increase in nanocrystal solubility with decreasing nanocrystal size can be described using the Gibbs-Thomson equation:[38]

$$S_r = S_b \exp(2\sigma V_m / rRT) \quad \text{Eq. (1)}$$

where S_r and S_b are the solubility of the nanocrystal and the corresponding bulk solid, σ is the specific surface energy, r is the radius of the nanocrystal, V_m is the molar volume of the materials, R is the gas constant, and T is the temperature. Since the temperature required for maintaining steady nanocrystal growth increases with increasing nanocrystal size, careful control of growth temperatures allows for accurate control of the average size and size distribution of the nanocrystals synthesized in a given reaction. Size and size distributions are monitored from the peak wavelengths and widths of the absorption or emission spectra. When the desired properties are attained, the temperature is reduced to prevent further growth or dissolution.

Variations to this organometallic approach have resulted in synthesis schemes that yield gram-quantities of high-quality qdots. In particular, Peng and coworkers[39, 40] have demonstrated that organometallic precursors (e.g. dimethyl cadmium) can be replaced with non-pyrophoric and less costly "greener" reagents (e.g. cadmium oxide, CdO, or cadmium acetate, $Cd(Ac)_2$). These "alternative routes" to the synthesis of qdots in organic media can been used to reproducibly prepare high-quality CdS, CdSe, and CdTe nanocrystals.[17, 39, 41] In addition, since nanocrystals formed from greener reagents exhibit slower reaction kinetics (e.g. slower nucleation), extended nucleation periods allow increased quantities of "greener" precursors to be injected at the start of the reaction. This is a promising approach to scaling up the synthesis of high quality qdots.

3.2. Capping of Quantum Dots

Due to the high surface-to-volume ratio of nanocrystals, the structural, optical, and electronic features of semiconductor qdots are heavily influenced by their surface properties. In effect, the surface properties of qdots are often manipulated to enhance the stability, processibility, and optical properties of these nanocrystals. The following sections will outline general strategies that are commonly used to manipulate the surface properties of qdots through the coating ("capping") of surfaces with organic ligands and inorganic materials.

3.2.1. Organic Capping

Coating of qdot surfaces with organic ligands typically facilitates the production of stable, high-quality qdots in organic media. A classic example of this effect is observed in the synthesis of CdSe qdots in hot solutions of amphiphilic tri-n-octylphosphine oxide (TOPO). In these solutions, the hydrophilic phosphine oxide groups coordinate to Cd sites on the qdot surface while the hydrophobic alkyl chains form a densely packed surface coating that stabilize nanocrystals against aggregation. The non-polar alkyl chains at the solid-liquid interface between the nanocrystals and solvent cause the resulting qdots to be soluble in organic solvents, such as chloroform and hexane. As a result, these TOPO-coated qdots can be selectively precipitated out of initial reaction mixtures containing excess ligands and unreacted precursors through the addition of polar solvents (e.g. methanol) and subsequent centrifugation. Since increasing the solvent polarity results in precipitation of smaller nanocrystals, repeated precipitation of qdot mixtures with increasing solvent polarities can be used to separate mixtures of quantum dots with different sizes; this process is referred to as "size-selective precipitation". The purified qdots typically have relative size distributions of less than 5%. Lastly, the capping of organic ligands on the qdot surfaces tends to improve their overall quantum yields.

3.2.2. Inorganic Capping

The fluorescence efficiency of qdots can be greatly enhanced by the capping of the nanocrystals inside an inorganic shell. For instance, monolayers of zinc sulphide (or cadmium sulphide) can be grown epitaxially on CdSe qdots by drop-wise injection of zinc (cadmium) and sulphur precursors into solutions of qdots at moderate temperatures (150-240°C).[16, 31] These reaction conditions favour deposition of the capping precursors onto the qdot surface over homogeneous ZnS (CdS) nucleation. The resulting core/shell nanocrystals exhibit luminescence quantum yields up to 85%, in contrast to the maximum quantum yields of ~15% observed for TOPO-capped CdSe qdots.[32]

In order for the capping layer to produce increased qdot quantum yields, the inorganic layer must be composed of a semiconductor with larger bandgap energy than the core qdot. In addition, for efficient capping, the bond length of the semiconductor comprising the capping layer must be similar to that of the core. The observed increase in quantum yield upon growth of an inorganic shell can be attributed to the removal of surface defects ("trap states") on the nanocrystals, a process referred to as electronic passivation. For CdSe nanocrystals, these surface defects typically result in broad emission at 700-800 nm, which are removed with the growth of an inorganic shell with a wider bandgap than the CdSe core. The removal of these trap states increases the number of electrons that undergo radiative recombination as described in Section 2 on Optical and Electronic Properties of Quantum Dots. Thus, the resulting nanocrystals exhibit an increase in fluorescence efficiency. Furthermore, capping with a semiconductor layer also prevents the photo-oxidation of the core qdot. Thus, while uncapped qdots may degrade within a month or two when stored in air at room temperature, long-term storage of capped qdots under similar conditions can often be achieved with minimal effect on their optical properties.

4. BIOLOGICAL APPLICATIONS OF QUANTUM DOTS

Over recent years there has been much excitement surrounding the potential applications of qdots to biological research, including the development of optical probes for biomedical imaging, bioassays and biosensors. For example, the novel size-tunable optical properties of qdots and universal conjugation strategies have generated a great impetus for the development of biological probes based on qdots.[7] In the coming sections, we will discuss some practical issues in developing qdots suitable for biological applications, and will review some proof-of-concept studies that highlight the unique advantages of qdots in biological research.

4.1. Biocompatible Quantum Dots

One of the major challenges to using qdots for developing biological probes and sensors is their surface chemistry. Since high-quality qdots are synthesized in organic solvents, such as TOPO, they are not water-soluble and, hence, incompatible with biological systems. In the last six years, great efforts have been placed on modifying the surface chemistry of qdots to render them biocompatible.[8, 9, 21, 42, 43] One approach to achieving qdot biocompatibility has been the use of surface exchange techniques, where TOPO molecules on the qdot surface are displaced by bifunctional molecules, such as mercaptoacetic acid and phospho-alcohols. One end of these bifunctional molecules contain functional groups (-SH or –P) that interact with metal atoms on the surface of the qdots and displace non-polar TOPO molecules. The other end of these bifunctional molecules typically contains polar alcohol or carboxylic acid functional groups, thus rendering the qdots extremely polar and water-soluble. Furthermore, the alcohol and carboxylic groups can react with biomolecules such as proteins, peptides, and oligonucleotides through several different synthetic techniques.

In a second approach to achieving qdot biocompatibility, molecules can be designed to interact with the TOPO molecules on the surface of qdots.[21, 43] These molecules typically contain both a hydrophobic and hydrophilic region (e.g., phospholipids). The hydrophobic end interacts with the TOPO molecule through hydrophobic-hydrophobic interactions, while the hydrophilic end, containing carboxylic acid or alcohol functional groups, protrudes from the surface of the quantum dot. The qdots can be locked into the organic-shell by cross-linking the surface-stabilizing molecule. This prevents the organic ligands from desorbing from the surface of the qdots and stabilizes the qdots against flocculation.

The functional groups provide a means to conjugate biomolecules (e.g., proteins, peptides, oligonucleotides) to form an optical probe. Conventional carbodiimide chemistry has been employed to catalyze this linkage (e.g., qdots containing carboxylic acids can react to biomolecules containing primary amines to form an amide bond). Electrostatic interactions can also be used to link biomolecules onto the surface of qdots. For example, Mattoussi and coworkers[44, 45] engineered a protein with a positive-charged leucine zipper and demonstrated the adsorption of such a protein onto the surface of a negatively-charged qdot.

4.2. Quantum Dots for Biomedical Imaging Applications

Biocompatible qdots have spurred great interest in their possible use as fluorescent labels for biological imaging applications. To date, qdots have been used to image cellular components (e.g. DNA, nuclear antigens,[21] and cytoplasmic components), cells,[9] and tissues[46] to name a few examples. Figure 4 shows a differential interference contrast (DIC) and fluorescence image of transferrin-coated quantum dots entering HeLa cells through receptor-mediated endocytosis.

One distinct advantage of using qdots over organic fluorophores for biological applications is the ability to perform multiplexed colour-coded imaging and detection, a general technique that uses multi-coloured fluorophore labels to simultaneously identify different biological targets and study the interactions between these targets.[7] Multiplexed imaging of qdots can be used to investigate different biological processes by tagging different biological targets with qdots of many different emission wavelengths and simultaneously exciting them with a single excitation wavelength as discussed in Section 2 on Optical and Electronic Properties of Semiconductor Quantum Dots. For instance, Mattoussi and coworkers used green, yellow, and red quantum dots to study the behavioural differences between starved and unstarved AX2 amoebae cells.[45] While the starved cells were shown to form aggregate centres (e.g. possibly via chemotaxis in response to signals sent by one another), the unstarved cells did not form such aggregates. In another study, Alivisatos and coworkers labelled mouse fibroblast cells using red and green coloured silica-coated CdSe-CdS qdots.[8] The red qdots were conjugated to biotin, which targets actin filaments, while the green qdots were conjugated to tri-methoxysilylpropyl urea and acetate groups, which bind electrostatically to the cell nucleus. Under both conventional wide-field and laser-scanning confocal fluorescence microscopes, the actin and the nucleus in the fibroblast samples were spatially and spectrally resolvable to the eye. Multiplexed imaging of qdots was further demonstrated by Wu and coworkers, who performed a series of experiments in which the nuclear antigens and microtubules in the cytoplasm were simultaneous stained and imaged using two different-coloured quantum dots.[21]

Several in situ studies have also investigated and exploited the resistance of qdots to photobleaching. In the above-mentioned experiments that probed AX2

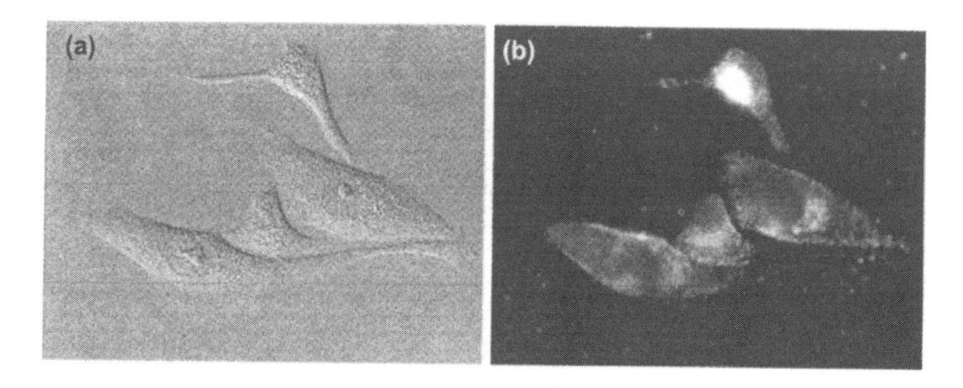

Figure 4. (a) DIC and (b) fluorescence image of transferrin-coated quantum dots entering HeLa cells. The HeLa cells are illuminated under optical excitation after endocytosing quantum dots. The fluorescence image was obtained using an Olympus epifluorescence microscope, 100 x, N.A. = 1.3, excitation filter = 480/40 and emission filter 535/50, and excited with a 100-W Hg lamp.[47]

amoebae cells, Mattoussi and coworkers observed no noticeable loss in quantum dot fluorescence intensities when monitoring cells under illumination for 14 hours by a 50 mW, 488 nm scanning continuous wave laser. Furthermore, no noticeable loss in fluorescence was observed when amoebae cells were visually tracked by taking 500 ms exposures once every minute for 14 hours using transmitted light from a halogen lamp and excitation light from a 75-W xenon lamp. Taking advantage of qdots photostability over such long terms, they were able to take time-lapse films of the cells undergoing aggregation and slug formation.[45] In another study, Dubertret and coworkers[43] injected qdots encapsulated in phospholipid block copolymer micelles into *Xenopus* embryos. Observations of micelle-qdots introduced to a single blastomere during very early cleavage stages indicated labelling of daughter cells, embryo, and subcellular structures with consistent fluorescent intensity until the tadpole stage of embryo development. These results suggest the suitability of qdots for use in lineage tracing studies in embryogenesis.[43] Furthermore, Derfus and coworkers[48] demonstrated the feasibility of using quantum dots labelled with epithelial growth factor to monitor cellular reorganization of hepatocytes for seven days. This work provides a means to study processes that are in important in the field of tissue engineering.

The bright luminescence and high photostability of single qdots allow for reliable single-molecule experiments, such as the labelling and tracking of single target biomolecules (e.g. proteins) in living cells.[49, 50] Dahan and coworkers[51] labelled individual glycine receptors (GlyRs) with qdots in order to study their lateral dynamics in the neuronal membrane, as well as their associations with boutons. They successfully recorded the paths of single GlyRs over more than 20 minutes, noting a vast improvement over the photostability of the organic Cy3 dye, which photobleached in about 5 seconds under similar conditions. They also obtained greater signal-to-noise ratio and spatial resolution using quantum dots in comparison to the Cy3 dye. This work opens the possibility of using qdot-bioconjugates to probe the interactions of individual biomolecules inside living cells in real-time.

Beyond cell imaging, quantum dot-bioconjugates have advanced toward *in vivo* animal imaging. Akerman and coworkers[46] coated the surfaces of qdots with targeting peptides. Upon intravascular injection in Balb/c mice, these qdot-bioconjugates were directed to the vasculatures of healthy cells, tissues, organs and breast tumours *in vivo*. After sacrificing the mice, tissues were obtained using conventional techniques and imaged to verify the accumulation of qdots in the targeted tissues. Recently, Nie and coworkers demonstrated the feasibility of performing whole-animal imaging by injecting quantum dots that target the folate receptor of tumours.[52] Similarly, Bawendi and coworkers[53] have shown the use of qdots for optically-guided surgery. In all of these studies, the use of qdots for *in vivo* applications was facilitated by the modification of the qdot surface chemistry. In the Akerman and co-workers study, the surface of the qdots was coated with polyethylene glycol to prevent the escape of qdots from the reticuloendothelial system, an animal's defense mechanism. Conversely, Bawendi and coworkers designed a thin organic coating on the surface of the qdots in order for the particles to freely traverse the lymphatic system.

4.3. Quantum Dots for Bioassays and Biosensors

In addition to cell and animal imaging, qdot-bioconjugates have been utilized for immunoassay-type applications. Chan and Nie used anti-immunoglobulin (IgG)-labelled qdots for detection of immunoglobulin (IgG) in a latex agglutination assay.[9] In this assay, anti-IgG-coated qdots were aggregated in the presence of IgG due to the formation of anti-IgG-to-IgG complex. Mattousi and coworkers[54] have also used antibody-labelled qdots to develop sandwich immunoassays that can be used to detect multiple analytes using the different qdot emission. In another approach, Pathak and coworkers[55] showed the use of qdots as an optical probe for fluorescence in situ hybridization.

The spectral flexibility and photostability of qdots also lend themselves to the development of robust and highly reusable fluorescent-based indicators or sensors. The idea of using quantum dots for the development of sensors based on fluorescence resonance energy transfer (FRET) has recently been explored by Mattoussi and coworkers, who designed and implemented two sugar sensors that employ qdots conjugated to engineered variants of the *Escherichia coli* maltose-binding protein (MBP) as FRET donors.[56] In the first design, an acceptor dye occupies all MBP saccharide binding sites located in close proximity to the qdot which quenches the qdot fluorescence by about 50%. Upon the addition of maltose, the acceptor dye is displaced from the sensor, thus restoring the maximum qdot fluorescent intensity. Based on similar principles, the second design used an intermediate fluorescent dye conjugated on the MBP to improve efficiency of energy transfer from the qdots to the acceptor dye in the saccharide binding site.

Recent studies have focused on the use of qdots for developing a multiplexed optical coding scheme for biomolecules (e.g. proteins, DNA).[57] In one approach, microbeads with unique optical signatures (e.g. fluorescence spectra) are created by infusing them with varying ratios of different-coloured qdots. Theoretically, qdots of 6 different colours and 10 different intensity levels can be used to generate microbeads with 10^6 unique optical signatures. Each of these optical signatures, or "barcodes", can be assigned to a unique biomolecule by tagging each microbead with a unique biomolecule (e.g. single-stranded DNA) that has specific binding affinity for a target biomolecule (e.g. complementary DNA strand). This barcoding technology may lead to developments in high-throughput genetic screening and medical diagnostics. In such applications, a target gene or protein would be detected by screening biological mixtures with libraries of qdot microbeads and subsequent single-bead spectroscopy for the detection of the optical signature of the analyte of interest.

5. FUTURE CONSIDERATIONS IN BIOLOGICAL APPLICATIONS OF SEMICONDUCTOR QUANTUM DOTS

The next decade will see a great increase in scientific research into the biological applications of semiconductor qdots. While proof-of-concept studies in research laboratories have demonstrated great promise in the use of qdots for biomedical imaging and detection, several issues will need to be addressed before qdots find their way to large-scale clinical application. In particular, researchers will need to study toxic and pharmacokinetic effects of qdots *in vivo*. Recent work by Derfus and coworkers[48] suggested that primary hepatocytes (i.e. liver cells that are responsible for detoxification of the blood) suffer from Cd-poisoning when interacting with CdSe qdots. As a result, advances in surface coatings (e.g. ZnS capping) and other strategies for minimizing qdot cytotoxicity will be necessary for the successful use of qdots in clinical applications, such as their use as contrast agents for molecular function imaging. In the meantime, biological research using quantum dots will continue to present new insights into biological processes[51] and to develop methods for the high-throughput screening of genes and proteins for drug discovery and disease detection.[46, 57] In addition, research will focus on the synthesis and self-assembly semiconductor nanocrystals into unique shapes (rods, tetrapods, helices) that may facilitate the development of multifunctional nanostructures for use in drug delivery and tissue engineering applications.[58-60]

6. ACKNOWLEDGMENTS

We thank Jonathan Lai and Mr. Fred Neub for help in preparing Figure 1, and Sawitri Mardyani and Wen Jiang for help in preparing Figure 4. A.S. acknowledges a fellowship from the National Sciences and Engineering Resource Council of Canada (NSERC). This work was supported by University of Toronto (Start-up grant), Connaught Foundation, Canadian Foundation for Innovation, Ontario Innovation Trust, and NSERC.

7. REFERENCES

1. C.L. Rieder and A. Khodjakov, Mitosis Through the Microscope: Advances in Seeing Inside Live Dividing Cells, *Science* **300** 91-96 (2003).
2. J. Lippincott-Schwarz and G.H. Patterson, Development and Use of Fluorescent Protein Markers in Living Cells, *Science* **300** 87-91 (2003).
3. J.M. Levsky, Shailesh M.Shenoy, R.C. Pezo, and R. H.Singer, Single-Cell Gene Expression Profiling, *Science* **297** 836-840 (2002).
4. S.F. Ibrahim and G.V.D. Engh, High speed cell sorting: fundamentals and recent advances, *Current Opinion in Biotechnology* **14** 5-12 (2003).
5. S.V. Nayak, A. S. Shivarudrappa, and A.S. Mukkamil, Role of Fluorescent Microscopy in Detecting Mycobacterium leprae in Tissue Sections, *Annals of Diagnostic Pathology* **7** (2), 78-81 (2003).

6. K. Truong, M. Gerbault-Seureau, M.-N. Guilly, P. Vielh, G. Zalcman, A. Livartowski, A. Chapelier, M.-F. Poupon, B. Dutrillaux, and B. Malfoy, Quantitative Fluorescence in Situ Hybridization in Lung Cancer as a Diagnostic Marker, *J Mol Diagn* 33-37 (1999).
7. W.C. Chan, D.J. Maxwell, X. Gao, R.E. Bailey, M. Han, and S. Nie, Luminescent quantum dots for multiplexed biological detection and imaging, *Current Opinion in Biotechnology* **13** (1), 40-46 (2002).
8. M. Bruchez, Jr., M. Moronne, P. Gin, S. Weiss, and A.P. Alivisatos, Semiconductor nanocrystals as fluorescent biological labels., *Science* **281** (5385), 2013-6 (1998).
9. W.C. Chan and S. Nie, Quantum dot bioconjugates for ultrasensitive nonisotopic detection., *Science* **281** (5385), 2016-2018 (1998).
10. S. Coe-Sullivan, W.-K. Woo, J.S. Steckel, M. Bawendi, and V. Bulovic, Tuning the performance of hybrid organic/inorganic quantum dot light-emitting devices, *Organic Electronics* **4** (2-3), 123-130 (2003).
11. S. Coe, W.-K. Woo, M. Bawendi, and V. Bulovic, Electroluminescence from single monolayers of nanocrystals in molecular organic devices, *Nature* **420** 800-803 (2002).
12. A.J. Nozik, Quantum dot solar cells, *Physica E* **14** 115 –120 (2002).
13. R.P. Raffaelle, S.L. Castro, A.F. Hepp, S.G. Bailey, and S. Issue, Quantum Dot Solar Cells, *Prog. Photovolt: Res. Appl.* **10** 433–439 (2002).
14. A. Singhal, H. Fischer, and W.C. Chan, *Unpublished data*. 2003.
15. J.E.B. Katari, V.L. Colvin, and A.P. Alivisatos, X-ray Photoelectron Spectroscopy of CdSe Nanocrystals with Applications to Studies of the Nanocrystal Surface, *J. Phys. Chem.* **98** 4109-4117 (1994).
16. B.O. Dabbousi, J. Rodriguez-Viejo, F.V. Mikulec, J.R. Heine, H. Mattoussi, R. Ober, K.F. Jensen, and A.M.G. Bawendi, (CdSe)ZnS Core-Shell Quantum Dots: Synthesis and Characterization of a Size Series of Highly Luminescent Nanocrystallites, *J. Phys. Chem. B* **101** 9463-9475 (1997).
17. R.E. Bailey and S. Nie, Alloyed Semiconductor Quantum Dots: Tuning the Optical Properties without Changing the Particle Size, *J. Am. Chem. Soc.* **125** (23), 7100-7106 (2003).
18. W.C.W. Chan. Ph.D. Thesis. "Semiconductor Quantum Dots for Ultrasensitive Biological Detection and Imaging." Indiana University, 2001.
19. A. Henglein, Small-particle research: physiochemical properties of extremely small colloidal metal and semiconductor particles, *Chemical Review* **89** 1861-1873 (1989).
20. C.A. Leatherdale, W.-K. Woo, F.V. Mikulec, and M.G. Bawendi, On the Absorption Cross Section of CdSe Nanocrystal Quantum Dots, *J. Phys. Chem. B* **106** 7619-7622 (2002).
21. X.E.A. Wu, Immunofluorescent labeling of cancer marker Her2 and other cellular targets with semiconductor quantum dots, *Nature Biotechnology* **21** 41-46 (2003).
22. A.P. Alivisatos, Perspectives on the Physical Chemistry of Semiconductor Nanocrystals, *J. Phys. Chem.* **100** 13226-13239 (1996).
23. C.B. Murray, C.R. Kagan, and M.G. Bawendi, Synthesis and Characterization of Monodisperse Nanocrystals and Close-Packed Nanocrystal Assemblies, *Annu. Rev. Mater. Sci.* **30** 545–610 (2000).
24. A.P. Alivisatos, Semiconductor clusters, nanocrystals, and quantum dots, *Science* **271** 933-937 (1996).
25. M. Nirmal, Brus, L. E., Luminescence photophysics in semiconductor nanocrystals, *Account Chemical Research* **32** 407-414 (1999).
26. A.A. Guzelian, J.E.B. Katari, A. Kadavanich, U. Banin, K. Hamad, E. Juban, A.P. Alivisatos, R.H. Wolters, C.C. Arnold, and J.R. Heath, Synthesis of size-selected, surface passivated InP nanocrystals, *J. Phys. Chem.* **100** 7212-7219 (1996).
27. J.A. Prieto, G. Armelles, J. Groenen, and R. Cales, Size and strain effects in the E-1-like optical transitions of InAs/InP self-assembled quantum dot structures., *Appl. Phys. Lett.* **74** 99-101 (1999).
28. L. Manna, E.C. Scher, L.-S. Li, and A.P. Alivisatos, Epitaxial Growth and Photochemical Annealing of Graded CdS/ZnS Shells on Colloidal CdSe Nanorods, *J. Am. Chem. Soc.* **124** (24), 7136-7145 (2002).
29. S.F. Wuister, I. Swart, F.V. Driel, S.G. Hickey, and C.D.M. Donega, Highly Luminescent Water-Soluble CdTe Quantum Dots, *Nano. Lett.* **3** (4), 503-507 (2003).
30. C.B. Murray, D.J. Norris, and M.G. Bawendi, Synthesis and Characterization of Nearly Monodisperse CdE (E = S, Se, Te) Semiconductor Nanocrystallites, *J. Am. Chem. Soc.* **115** 8706-8715 (1993).
31. M.A. Hines and P. Guyot-Sionnest, Synthesis and Characterization of Strongly Luminescing ZnS-Capped CdSe Nanocrystals, *J. Phys. Chem. B* **100** 468-471 (1996).
32. X. Peng, M.C. Schlamp, A.V. Kadavanich, and A.P. Alivisatos, Epitaxial Growth of Highly Luminescent CdSe/CdS Core/Shell Nanocrystals with Photostability and Electronic Accessibility, *J. Am. Chem. Soc.* **119** 7019-7029 (1997).
33. A.R. Kortan, R. Hull, R.L. Opila, B.M. G., M. Steigerwald, P.J. Carroll, and L. Brus, Nucleation and Growth of CdSe on ZnS quantum crystallite seeds, and vice versa, in inverse micelle media, *J. Am. Chem. Soc.* **112** 1327-1332 (1990).

34. T. Dannhauser, M. O'neil, K. Johansson, D. Whitten, and G. Mclendon, Photophysics of quantized colloidal semiconductors: dramatic luminescence enhancement by binding of simple amines., *J . Phys. Chem.* **90** 6074-6076 (1986).
35. C. Dameron, R.N. Reese, R.K. Mehra, A.R. Kortan, P.J. Carroll, M. Steigerwald, L. Brus, and D.R. Winger, Biosynthesis of cadmium sulfide quantum semiconductor crystallites, *Nature* **338** 596-597 (1989).
36. C. Dameron, B. Smith, and D. Winge, Glutathione-coated cadmium-sulfide crystallites in Candida glabrata, *J. Biol. Chem.* **264** 17355 - 17360 (1989).
37. V.K. Lamer and R.H. Dinegar, Theory, Production and Mechanism of Formation of Monodispersed Hydrosols, *J. Am. Chem. Soc.* **72** (11), 4847-4854 (1950).
38. T. Sugimoto, Preparation of monodispersed colloidal particles, *Advances in Colloid and Interface Science* **28** 65-108 (1987).
39. L. Qu, Z.A. Peng, and X. Peng, Alternative Routes toward High Quality CdSe Nanocrystals, *Nanoletters* **1** (6), 333-337 (2001).
40. Z.A. Peng and X. Peng, Formation of High-Quality CdTe, CdSe, and CdS Nanocrystals Using CdO as a Precursor, *J. Am. Chem. Soc.* **123** 183-184 (2001).
41. I. Mekis, D.V. Talapin, A. Kornowski, M. Haase, and H. Weller, One-Pot Synthesis of Highly Luminescent CdSe/CdS Core-Shell Nanocrystals via Organometallic and "Greener" Chemical Approaches, *J. Phys. Chem. B* **107** 7454-7462 (2003).
42. H. Mattoussi, J.M. Mauro, E.R. Goldman, T.M. Green, G.P. Anderson, V.C. Sundar, and M.G. Bawendi, Bioconjugation of Highly Luminescent Colloidal CdSe-ZnS Quantum Dots with an Engineered Two-Domain Recombinant Protein., *Physica Status Solidi (b)* **244** (1), 277-283 (2001).
43. B. Dubertret, P. Skourides, S.J. Norris, V. Noireaux, A.H. Brivanlou, and A. Libchaber, In Vivo Imaging of Quantum Dots Encapsulated in Phospholipid Micelles, *Science* **298** 1759-1762 (2002).
44. H. Mattoussi, J.M. Mauro, E.R. Goldman, G.P. Anderson, V.C. Sundar, F.V. Mikulec, and M.G. Bawendi, Self-Assembly of CdSe-ZnS Quantum Dot Bioconjugates Using an Engineered Recombinant Protein, *J. Am. Chem. Soc.* **122** 12142-12150 (2000).
45. J.K. Jaiswal, H. Mattoussi, J.M. Mauro, and S.M. Simon, Long-term multiple color imaging of live cells using quantum dot bioconjugates.[comment], *Nature Biotechnology* **21** (1), 47-51 (2003).
46. M.E. Akerman, W.C. Chan, P. Laakkonen, S.N. Bhatia, and E. Ruoslahti, Nanocrystal targeting in vivo, *P.N.A.S.* **99** (20), 12617-21 (2002).
47. S. Mardyani, W. Jiang, H. Fischer, A. Singhal, and W.C. Chan, unpublished data, (2004).
48. A.M. Derfus, W.C.W. Chan, and S.N. Bhatia, Probing the Cytotoxicity of Semiconductor Quantum Dots, *Nano Lett.* (2003, in print.).
49. A.J. Sutherland, Quantum dots as luminescent probes in biological systems., *Current Opinion in Solid State and Material Science* **6** (4), 365-370 (2002).
50. M. Dahan, T. Laurence, F. Pinaud, and D. Chemla, Time-gated biological imaging by use of colloidal quantum dots, *Optics Letters* **26** (11), 825-827 (2001).
51. M. Dahan, S. Levi, C. Luccardini, P. Rostaing, B. Riveau, and A. Triller, Diffusion Dynamics of Glycine Receptors Revealed by Single-Quantum Dot Tracking, *Science* **302** 442-445 (2003).
52. C. Seydel, Quantum Dots Get Wet, *Science April 4* **300** 80-81 (2003).
53. S. Kim, Y.T. Lim, E.G. Soltesz, A.M. De Grand, J. Lee, A. Nakayama, J.A. Parker, T. Mihaljevic, R.G. Laurence, D.M. Dor, L.H. Cohn, M.G. Bawendi, and J.V. Frangioni, Near-infrared fluorescent type II quantum dots for sentinel lymph node mapping, *Nature Biotechnology* **22** (1), 93-97 (2004).
54. E.R. Goldman, G.P. Anderson, P.T. Tran, and H. Mattoussi, Conjugation of Luminescent Quantum Dots with Antibodies Using an Engineered Adaptor Protein To Provide New Reagents for Fluoroimmunoassays, *Anal. Chem.* **74** 841-847 (2002).
55. S. Pathak, Choi, S. K., Arnheim, N., Thompson, M. E., Hydroxylated quantum dots as luminescent probes for in situ hybridization, *Journal of the American Chemical Society* **123** 4103-4104 (2001).
56. I.L. Medintz, A.R. Clapp, H. Mattoussi, E.R. Goldman, B. Fisher, and J.M. Mauro, Self-assembled nanoscale biosensors based on quantum dot FRET donors, *Nature Materials* **2** 630-638 (2003).
57. M. Han, X. Gao, J.Z. Su, and S. Nie, Quantum-dot-tagged microbeads for multiplexed optical coding of biomolecules, *Nature Biotechnology* **19** 631 - 635 (2001).
58. L. Manna, E.C. Scher, and A.P. Alivisatos, Synthesis of Soluble and Processable Rod-, Arrow-, Teardrop-, and Tetrapod-Shaped CdSe Nanocrystals, *J. Am. Chem. Soc.* **122** (51), 12700-12706 (2000).
59. L.-S. Li, J. Hu, W. Yang, and A.P. Alivisatos, Band Gap Variation of Size- and Shape-Controlled Colloidal CdSe Quantum Rods, *Nano Lett.* **1** (7), 349-351 (2001).
60. E.D. Sone, E.R. Zubarev, and S.I. Stupp, Semiconductor Nanohelices Templated by Supramolecular Ribbons, *Angew. Chem. Int. Ed.* **41** (10), 1705 (2002).

POTENTIAL APPLICATIONS OF CARBON NANOTUBES IN BIOENGINEERING

Akil Sethuraman, Michael A. Stroscio, and Mitra Dutta*

1. INTRODUCTION

Nanotechnology deals with the design and manufacture of devices and structures with nanoscale features. These nanoscale features generally vary from about 0.1 to 100 nm (1 nm = 10^{-9} m). Nanoscience and nanotechnology portend numerous applications in the fields of biotechnology, biomedical engineering and electronics. Some of the common nanobiological applications of nanoparticles in the above-mentioned fields are encapsulation, DNA transfection, sensing and drug delivery.[1]

This chapter highlights potential applications of carbon nanotubes (CNTs) in the field of biomedical engineering. This discussion starts with a description of some of the basic properties of carbon nanotubes. Succeeding sections enumerate the interesting properties of these tubes and their applications in various fields.

Recently, functionalization of the walls of CNTs by peptides and other chemical agents has resulted in these tubes being bound to various biological entities. This has proved to be a major breakthrough and has resulted in numerous nanobiological applications using CNTs. In a related topic, the solubility of carbon nanotubes plays an important role in their purification and modification. Various techniques to increase the solubility of these tubes in organic and inorganic solvents have been discussed; these techniques will be highlighted herein. Moreover, a brief description on the conduction properties of these tubes as well as the advantages of combining nanotechnology with MEMS are included. Finally, the most recent advances in the field of carbon-nanotube technology and the applications of these tubes in various other fields are highlighted in this review. Recent carbon nanotube research has addressed issues regarding the toxicity of CNTs and the separation of semiconducting tubes from their conducting counterparts. There have been many recent advances in CNT technology that make the field of carbon

---

* Akil Sethuraman, Dept. of Bioengineering, Univ of IL at Chicago, Chicago, IL 60607. Michael A. Stroscio, Depts. of Bioengineering, Electrical and Computer Engineering, and Physics, Univ of IL at Chicago, Chicago, IL 60607. Mitra Dutta, Depts. of Electrical and Computer Engineering, and Physics, Univ of IL at Chicago, Chicago, IL 60607.

nanotechnology extremely potent. This review enumerates promising discoveries underlying the application of CNT technology to bioengineering.

2. CARBON NANOTUBES: BASIC PROPERTIES OF INTEREST IN THIS REVIEW

Nanotechnology refers to the manufacture and usage of devices and structures on the nanometer scale. The fact that these dimensional scales correspond approximately to those of molecular structures makes this a field rich in exploitable physical phenomena and opens the way to many potential applications that have not been possible, or even envisioned, in the past. Graphene is composed of carbon atoms organized in a 2-dimensional hexagonal lattice as shown in Figure 1. Single-wall carbon nanotubes are cylindrical structures that have the appearance of rolled sheets of graphene. These tubes, known for their high mechanical strength and thermal conductivity are also unique for their controllable properties as semiconductors and metals. The high strength results from the extremely strong carbon-carbon bond. As will be discussed in detail, there are many potential applications of these tubes are in the fields of electrical, mechanical and biomedical engineering.

The charge transport properties of CNTs make them suitable for applications as a variety of biosensors. To understand the potential of CNTs in biosensor applications, it is necessary to consider the electronic properties of CNTs. CNTs, owing to their high strength can be used in structural and mechanical applications. Indeed, the Young's modulus and tensile strength of a single-walled CNT are approximately 1050 GPa and 150 GPa, respectively. These values take on additional significance when compared with the corresponding values for steel, 208 GPa and 0.4 GPa, respectively. CNTs maybe single walled or multi walled depending on the number of sheets of graphene composing the CNT.

Nanotubes are classified as armchair, zigzag and chiral depending on how the sheets are rolled to form the cylinder. The different structures may be explained with the aid of a parameter called the chiral vector, C_h, which is defined by the following relation:

$$C_h = na_1 + ma_2$$

where n, m are integers, and a_1 and a_2 are unit vectors as shown in Figure 2. The radius of a carbon nanotube, R, is given by the following relation.[35]

$$R = C_h/2\pi = (\sqrt{3}d/2\pi)\,(n^2 + m^2 + nm)^{1/2}$$

where d is the carbon-carbon bond length having the value of 1.42 Angstroms. The chiral angle, θ, is the angle between the chiral vector and the unit vector a_1. The cosine of the chiral angle θ is given by,

$$\cos\theta = C_h\, a_1 / |C_h|\,|a_1| = 2_{2n+m}/2\,(n^2 + m^2+nm)^{1/2}$$

The so-called armchair CNTs are those having a chiral angle of 30° and the integers n and m have the same value. Zigzag CNTs are obtained when m = 0 corresponding to a chiral angle with value zero.[2]

In summary, the integers n and m define the orientation and the size of the tube while the length of the chiral vector is related to the circumference of the nanotube. These parameters may be determined using the techniques of STM (scanning tunneling microscopy) and TEM (transmission electron microscopy). [2]

CNTs have dramatically different electrical properties depending on the values of m and n. Specifically, CNTs behave as metallic structures when m and n differ by an integer multiplied by a factor of 3. In particular, if m and n differ by 0, 3, 6, etc., the CNT behaves as a metallic structure. Otherwise, the CNT behaves as a semiconductor. In the case of a semiconductor, there is a gap in the available allowed energy states, E_g, known as the band gap. In the absence of thermally excitation electrons, the band gap of a pure semiconductor represents an energy gap between the highest occupied energy level and the lowest unoccupied energy level. For chiral single-walled nanotubes, this energy gap is given approximately by $E_g/t = 100/d_t$, where t = 2.5 eV and d_t is the CNT diameter in Angstroms. One of the important properties of a CNT results from the nature of the chemical bonding in a graphene sheet. In a tetrahedral diamond lattice, the carbon atoms are bound in an sp^3 configuration. For the graphene-based CNTs, the bonding has an sp^2 hybridization and π-orbitals project radially outward in the direction normal to the CNT cylindrical surface at the location of each carbon atom. As will be discussed, these π-orbitals play a major role in efforts to chemically functionalize CNTs. The synthesis of carbon nanotubes usually results in the production of both metallic and semiconducting nanotubes of a variety of chiralities and of different diameters. This mixture needs to be sorted out so that they are appropriate for each anticipated application. Current research focuses on the techniques used to achieve this separation. As an example of one approach, this may be achieved by the oxidation of metallic tubes wherein the oxygen atom in air reacts with the carbon atoms to form an oxide. As discussed previously, nanotubes may be metallic or semiconducting depending on the values of m and n. Other factors that determine the nature of tubes are the number and radii of the graphene sheets making up the wall of the nanotube. Nanotubes with a single layer of graphene as their wall are known as single-walled nanotubes (SWNTs) and those with multiple graphene layers are known as multi-walled nanotubes (MWNTs). The absence of energy gaps in the armchair tubes is in accord with their metallic character while the presence of energy gaps in the zigzag tubes occurs because they are semiconducting. Some of the common techniques used to manufacture CNTs are the arc-discharge method, the laser ablation method and the catalytic technique that involves the deposition of hydrocarbons on transition metal catalysts. [3]

3. POTENTIAL APPLICATIONS OF CARBON NANOTUBES FOR DRUG DELIVERY

Encapsulation and the controlled release of drugs using nanospheres have proved extremely successful.[1,31] The drug to be released is surrounded by a polymeric biodegradable vehicle that is spherical in shape. As the spheres degrade, the drugs are

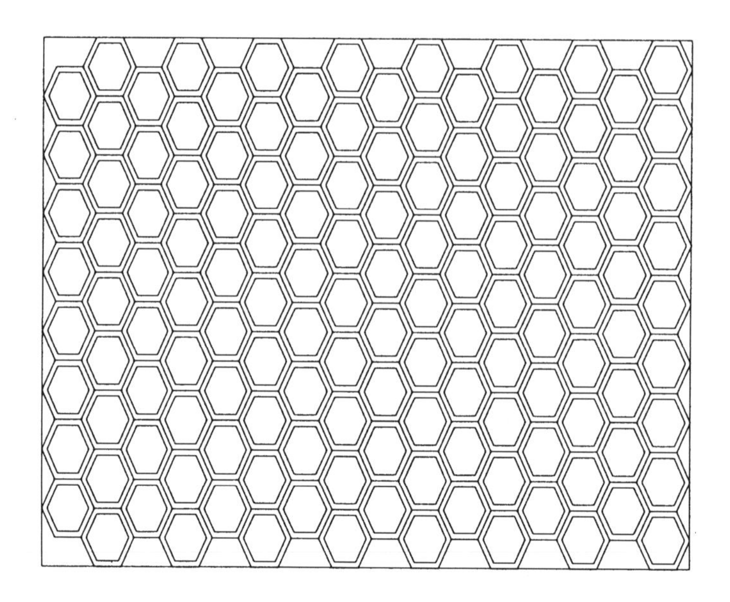

Figure 1. A single layer of graphite, graphene is composed of a hexagonal array of carbon atoms.

released into the system. By immobilizing specific peptide sequences onto the surface of the nanosphere, the permeation probability of these spheres through cell membranes is enhanced. The process of incorporating DNA into living cells is referred to as DNA transfection. DNA, being anionic in nature binds to the outer surface of nanospheres coated with positively charged ammonium groups. The bound DNA is then introduced into the cells. The process of DNA transfection is also being carried out using nano liposomes (membrane bound vesicles with a lipid bilayer).[34] Spherical particles, owing to their ease of manufacture, have been the obvious choice in the above-mentioned applications. Another alternative is to use nanotubes. It may be possible to load nanotubes with a desired material. The functionalization of the inner and outer surfaces of these tubes with various chemicals is also a possible approach. Gold nanoparticles possess an extremely high absorption coefficient that makes them ideal visual indicating agents.[32] These particles exhibit various colors depending on the size and the shape of the particle. The micro and nano-tubes, which have evoked interest among researchers are organosilicon polymer tubes, lipid microtubes, carbon nanotubes, peptide nanotubes and template-synthesized nanotubes. By attaching functional groups to tube sidewalls, nano- tubes can be used in processes like extraction and catalysis. By immobilizing antibodies, nanotubes can also be used to separate enantiomers from a racemic mixture. Enantiomers are usually difficult to separate largely due to their chemical similarity.[33] The process of separation could be made easier by the use of side-wall-functionalized nanotubes.

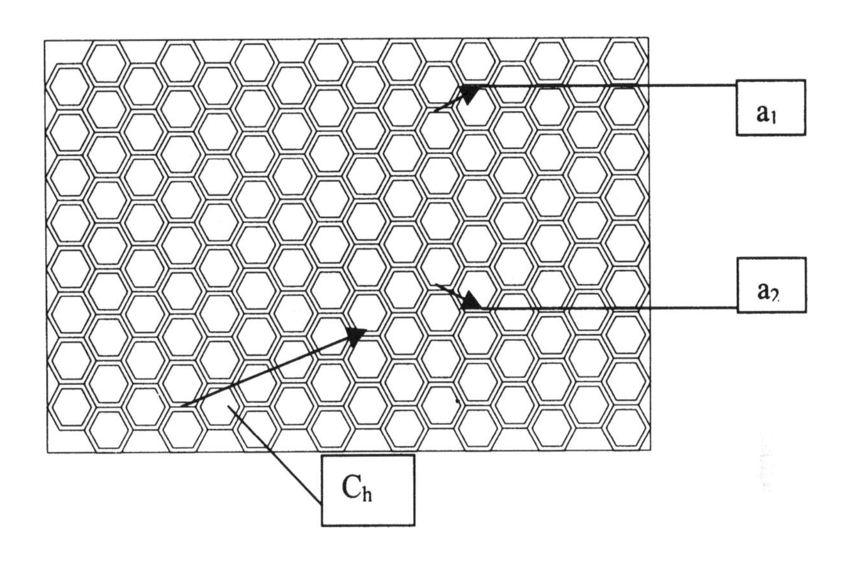

Figure 2. Chiral vector of a CNT and its unit vector.

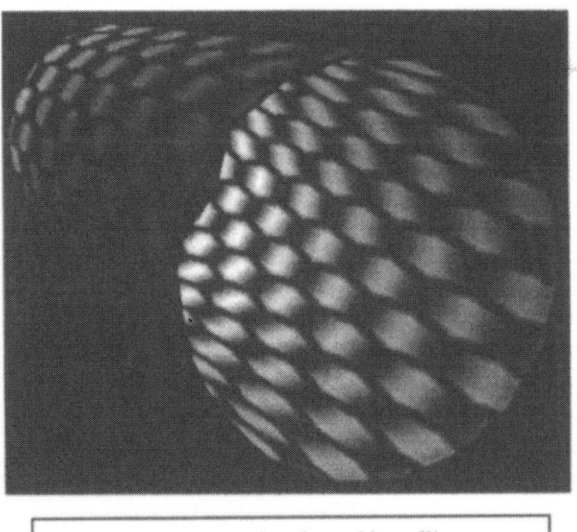

Carbon nanotubes formed by rolling graphene sheets

Figure 3. Image of a single-walled carbon nanotube (SWNT).

Figure 4. Image of a multi-walled carbon nanotubes (MWNT).

4. CHEMICAL FUNCTIONALIZATION OF CARBON NANOTUBES

The small size, high strength and interesting electrical and mechanical properties of carbon nanotubes make them potentially useful in the field of bioengineering. This section addresses potential applications of nanotubes in neural engineering. As will be discussed, carbon nanotubes (CNTs) are potentially useful as biosensors that record electrical activity in neuronal segments.

The current methods adopted to record neuronal electrical activity are not well-suited to monitoring neuronal activity on a neuron at multiple locations determined with nanoscale precision. CNTs due to their extremely small size may be useful in such accurate measurements since these tubes can be placed at sites where electrical changes take place. In order to bind nanotubes to neurons, it is necessary to functionalize the walls of the tubes. This functionalization procedure is then followed by the immobilization of peptides or proteins on functionalized nanotube surfaces. The proteins selected must be such that they bind to neurons. The use of bi-functional molecules like 1-pyrene butanoic acid, succinimidyl ester as a crosslinker between the nanotube and the protein has been investigated.[4] The pyrenyl group of this compound reacts strongly with the graphite plane via π-stacking (overlap of π bonds between aromatic side chains). Attachment by π-stacking preserves both the structure and the electronic properties of nanotubes. The ester part of this compound binds to the amine group, thus acting as a bridge between the protein and the CNT. Chen et al.[4] have carried out functionalization by incubating CNT samples in a solution of 1-pyrene butanoic acid, succinimidyl ester. The excess reagent was then washed away by rinsing with DMF (dimethyl formamide). Proteins used for immobilization were ferritin, streptavidin and biotin-PEO-amine. Immobilization was achieved by incubation of the sample in protein solution at room temperature for 18 hours. Samples were then rinsed with water and dried. The analytical techniques of XPS (X-Ray photoelectron spectroscopy) and TEM (transmission electron microscopy) indicated that the proteins were immobilized on the CNTs. No protein attachment was observed in the case where the CNTs were not functionalized with 1-pyrene butanic acid, succinimidyl ester. In a later section of this review, the use of the structures of Figure 6

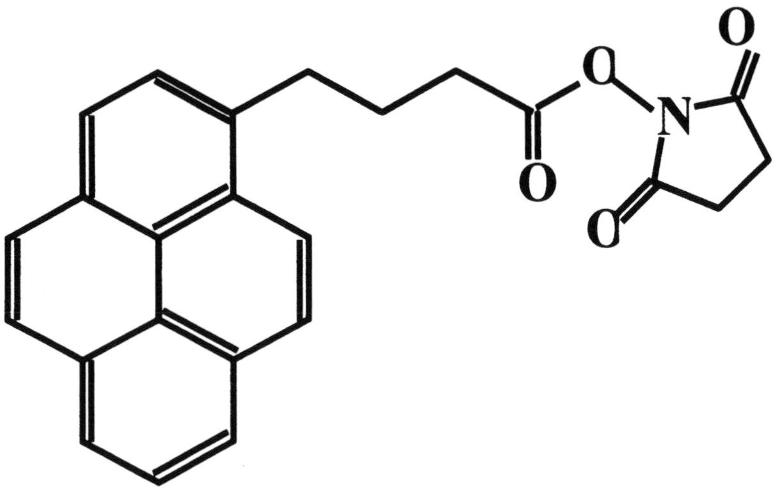

Figure 5. The pyrenyl group in 1-pyrenebutanic acid, succinimidyl ester is shown to the left.

for binding CNTs with nanoscale spatial precision to neurons will be discussed. As mentioned previously, CNTs portend applications as biosensors. In many of these applications, the binding of a target molecule to the CNT will modify the electrical states on the surface of the CNT and a subsequent change in the conductivity of the CNT results. The binding of unwanted proteins onto the nanotube surface is referred to as non-specific binding (NSB). The non-specific binding of proteins is highly undesirable since it may modify the electrical conductivity of nanotubes.

Single walled carbon nanotubes (SWNTs), due to their electrical sensing capabilities may be used potentially to detect proteins and other biological molecules in fluids. In order to use these tubes as sensors, the NSB of proteins has to be eliminated. Chen et al.[5] have suggested methods to reduce the nonspecific binding of proteins and increase binding affinities of proteins of interest. The NSB was demonstrated using AFM (atomic force microscopy) and QCM (quartz crystal microbalance). Some of the proteins used by Chen et al.[5] were Biotin and SpA. The NSB of proteins may be attributed to various hydrophobic interactions between the protein and the surface of the nanotube. To increase the protein resistance of the nanotube, compounds containing PEO (poly ethylene oxide) subunits were attached to nanotube surfaces. Tween 20 and pluronic triblock copolymers proved to be the most successful PEO compounds in imparting high resistance to proteins. Attachment of these compounds offers a 2-fold advantage. Apart from increasing protein resistance, they also help in increasing the water solubilization of nanotubes. Strong adherence of these compounds and a marked reduction in NSB of proteins were confirmed using AFM and QCM. Since the conductance level of semiconducting nanotubes are more sensitive to electric field changes when compared to metallic nanotubes, a higher percentage of the tubes manufactured were semiconducting in nature. Before NSB, the conductance level of tubes was much higher. A marked decrease in conductance was observed regardless of whether the protein was positively or negatively charged. Chen et al.[5] also demonstrated that this method could be used to

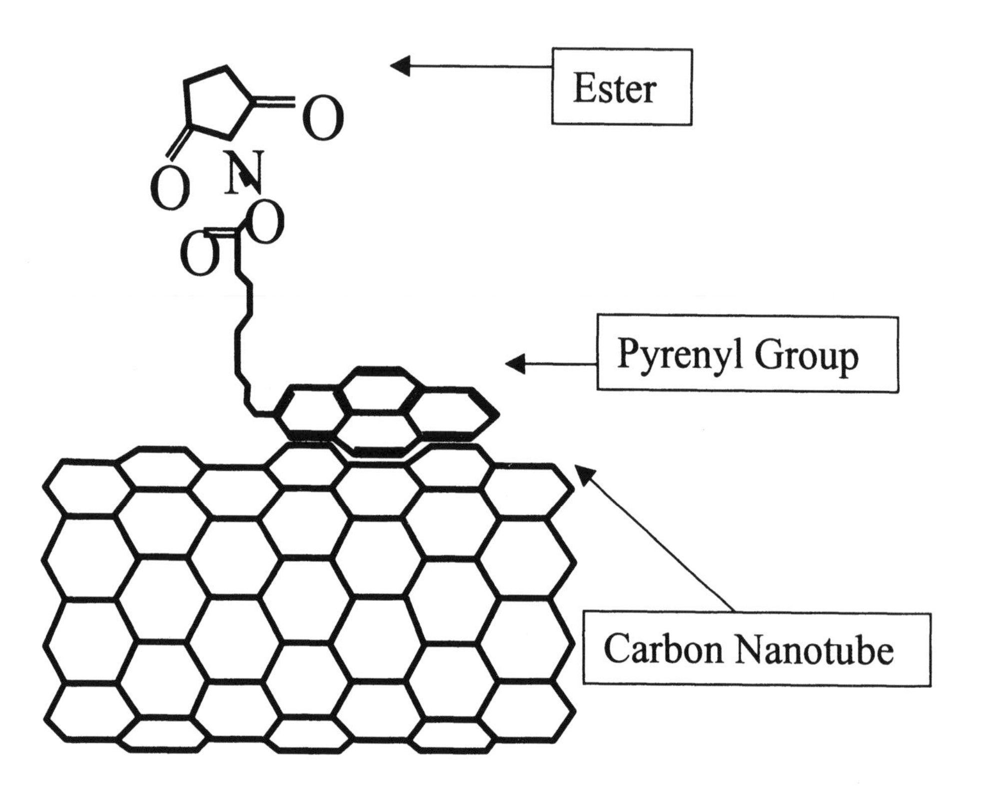

Figure 6. CNT functionalized with 1-pyrenebutanic acid, succinimidyl ester.

identify antigens and antibodies. Antigens bound to the nanotubes retained their activity and did bind to their respective antibodies. The use of nanotubes as sensors eliminates the labeling procedure that is usually performed to identify the binding of antibodies to their specific antigens. As biosensors, these nanotubes could prove to be extremely powerful tools in the fields of nanobiotechnology and proteomics.

Shim et al.[6] have studied the interaction between streptavidin/biotin system with carbon nanotubes. Shim et al. established that streptavidin binds nonspecifically onto CNTs due to hydrophobic interactions and that this can be reduced significantly by coating nanotubes with a surfactant and PEG (protein resistant polymer). The Triton –

PEG combination proved to be effective in reducing NSB of proteins. The surfactant also helped in the increased adsorption of PEG onto nanotube surfaces. In Reference 6, the process of adsorption was followed by the addition of biotin via the amine terminated PEG chains. When immersed in streptavidin solution, the nanotubes showed increased selective adhesion of the protein all along the tube length. In addition, the binding of fibrinogen on nanotube surfaces was studied and results indicated that small molecules have a much higher adsorption capacity than large ones owing to the covalent nature of interactions and the presence of large areas for interaction between small molecules and SWNTs. The effectiveness of any surfactant-polymer system in reducing NSB of proteins depends on the coverage and uniformity of the adsorbed polymer layer. In related research, the immobilization of various metalloproteins and enzymes are being carried out on pure, oxidized nanotubes. The binding of glucose oxidase onto SWNTs is one such example of enzyme immobilization. The corresponding nanotubes images were characterized using AFM.[7]

The unique properties of CNTs have been exploited in fields ranging from electronics to biotechnology. As discussed recently by Mattson et al., one of the most recent applications of these tubes has been in the field of neural engineering. Furthermore, Mattson et al.[8] have demonstrated that CNTs can be used as substrates for neuronal growth by the functionalization of tubes with certain bioactive molecules. The extension of neurites and the formation of synapses are controlled by a highly specialized region at the tip of a neurite called the growth cone. In spite of the isolation of various neurotrophic factors and neurotransmitters responsible for the growth and inhibition of neurons, the mechanisms responsible for neurite outgrowth on the nanoscale have not been established. The use of MWNTs as substrates for neuronal growth has been studied using embryonic rat hippocampal neurons. The nanotubes dispersed in ethanol were functionalized with 4-hydroxynonenal (4-HNE) by incubating the tubes in an acidic solution of ethanol and 4-HNE. SEM images confirmed neuronal attachment on nanotubes and neurite formation. It was also seen that the nanotubes played no part in influencing the direction of neurite extension. The attachment of 4-HNE all along the nanotube length was confirmed by the use of a 4-HNE antibody. Cells seeded onto unmodified nanotubes resulted in neurite extension but no branching was observed. This led to the conclusion that the growth cones were weakly bound to unmodified nanotube surfaces. An increase in neuronal outgrowth and branching on 4-HNE bound CNTs confirmed the role played by 4-HNE. Some of the parameters that were quantified during this study were the number of neurites per cell, the neurite length and the number of branches per neurite.

The current methods employ flat substrates for neuronal growth. These substrates bear no resemblance to the external environment encountered by cells *in vivo*. The use of functionalized nanotubes has proved successful, as it helps in neurite extension as well as branching. Also, the CNT diameters may be chosen to be similar to those of neurites, thus resulting in more molecular interactions between the tubes and neurons.

Williams et al.,[9] have shown recently that CNTs may be used as sensors for detecting biological molecules. In these studies, SWNTs were used since they are compatible dimensionally with various biological molecules used in the experiment. First, SWNT ropes were shortened using a mixture of sulfuric and nitric acid. This was followed by the introduction of carboxyl groups using 1M hydrochloric acid. The carboxyl end groups were then converted into ester linkages with the aid of NHS (N-hydroxysuccinimide) to form SWNT-NHS esters. Subsequent reaction of these tubes with PNA (Peptide Nucleic

Acid; NH_2-Glu-GTGCTCATGGTG-$CONH_2$) resulted in the formation of PNA bound SWNTs. DMF (dimethyl formamide) was the solvent used in the above reaction. DNA sequences, with bases complementary to those present in PNA were prepared from double stranded DNA using restriction enzymes. The cleaved products were then joined to form single stranded nucleotides. These were then attached to the PNA bound SWNTs and images taken using AFM. It was also shown that the DNA binding was predominant at the tube ends. The reasons for choosing PNA as the cross linker were its high compatibility with solvents, high resistance to enzymatic degradation and the thermal stability of the PNA-DNA pairs.[9]

5. SOLUBILIZATION OF CARBON NANOTUBES

Increasing the solubility of carbon nanotubes in water facilitates chemical modification, purification and separation of nanotubes from insoluble impurities. O'Connell et al. discuss a novel polymer wrapping technique to increase the solubility of CNTs. Some of the linear water-soluble polymers used for this purpose are PVP (poly vinyl pyrrolidone) and PSS (polystyrene sulfonate). In Ref. 10, single walled nanotubes are dispersed in a solution of SDS (sodium dodecyl sulphate) and the PVP polymer was added to the mixture that was incubated for 12 hours at around 50 degrees centigrade. PVP-wrapped CNTs were obtained in this process and the excess polymer and SDS were removed by high speed centrifugation. The adhesion strength of polymer was tested using the technique of field flow fractionation (FFF) and AFM results showed strong uniform wrapping of the polymer onto the carbon nanotubes. Quantification of the amount of polymer in solution was accomplished using NMR spectroscopy and the total polymer concentration was obtained using absorption spectroscopy methods. This difference in concentration gives the amount of polymer wrapped around the nanotubes. Thermodynamic factors associated with the wrapping procedure were considered and helical wrapping of the polymer around the nanotubes was suggested as the possible occurring mechanism.[10]

As discussed previously, nanotubes have been made soluble in water by wrapping polymers around them. By covalently attaching alkanes onto the nanotube surfaces, they also may be made soluble in organic solvents like chloroform, methylene chloride and tetrahydrofuran. Alkylation of nanotubes using two different mechanisms and the subsequent removal of bound alkanes from tube sidewalls has been carried out successfully using single walled fullerene nanotubes.

Indeed, Mickelson et al., fluorinated nanotubes using elemental fluorine to yield fluorinated nanotubes prior to alkylation. In the first technique, alkylation was carried out using alkyllitium species (methyl to dodecyl). In the experiments of Boul et al.,[12] hexylated nanotubes were produced using a solution of hexyl lithium in hexane and ethanol was used to remove any excess reagent. The second technique employed the use of Grignard reagent (alkyl magnesium bromides in tertahyrofuran). The fluorinated tubes could be stripped off fluorine with the aid of hydrazine that removes the fluorine layer to yield pure nanotubes. Similarly, pure tubes were obtained from alkylated tubes by heating the tubes in air for one hour at 250 degrees centigrade (oxidation). AFM images of tubes before and after oxidation indicated that the nanotubes were much thicker before the oxidation process was carried out. This was further confirmed by measuring the electrical resistance of pure and alkylated tubes. The resistance of alkylated tubes was much higher than those of the pure tubes (test for alkylation reversibility). Also, no shortening of tubes were observed. Regarding the alkylation process, Boul et al.[12] also addressed the question of whether the functionalization was by chemisorption or physisorption. UV-VIS spectra of pure, fluorinated and alkylated nanotubes suggested chemisorption of the alkyl species. In summary, it was demonstrated that alkylated tubes were soluble in chloroform, THF and methylene chloride, insoluble in solvents including hexane and toluene.

Solubilizing nanotubes facilitates the purification and separation of nanotubes. The current methods of solubilization of nanotubes involve the use of synthetic polymers. A drawback in using these polymers is that they are not very biocompatible. Natural polymers owing to their biocompatibility may be used to wrap nanotubes. Reference 13 considers the use of starch and other natural substances like gum arabic and glucosamine to dissolve carbon nanotubes. In particular, carbon nanotubes failed to dissolve when in contact with an aqueous solution of starch. However, the nanotubes dissolved when an aqueous solution of starch-iodine complex was used. This was attributed to the fact that the amylose component of starch combined with the iodine molecules to form a helix that coils around the tubes. It was also established that amylose (the linear component of starch) was the main component which helps dissolve the nanotubes while amylopectin (the branched portion) increases the solubilizing capacity of starch wrapped CNTs. Samples observed under an atomic force microscope revealed clusters of nanotubes wrapped with starch. The nanotubes may be precipitated from solution by the addition of saliva to the mixture. Amylase present in saliva helps break the amylose chains and precipitates the tubes from solution. The mechanical and electrical properties of individual carbon tubes are far superior to those of ropes. Gum arabic, a glycopolymer, has been used to isolate individual tubes from ropes. CNTs are known to have great affinity for amine groups. Compounds that are highly soluble in water and that possess amine groups could help increase the solubility of CNTs. One such compound that has been tested with nanotubes is glucosamine. Clearly, potential applications of these nanotubes[13] are in the field of targeted drug delivery wherein nanotubes with antibodies grafted onto them can be used to target and destroy tumor cells.

6. BINDING PROTEINS TO NEURONS

Proteins with the SIKVAV (serine-isoleucine-lysine-valine-alanine-valine) sequence are known to bind to neurons.[14] Laminin-1, the basement membrane protein stimulates formation of outgrowths from neuronal cells and promotes cell adhesion in specific cell lines. The IKVAV and LQVQLSIR sequences were found to be the two major outgrowth-promoting sites in the Laminin-1 chain. This fact has been demonstrated using Laminin-1 peptides and cells isolated from the cerebellar cortex of mice. Cell adhesion has been observed with other peptides but the rate of neurite outgrowth production was far less than that seen with Laminin-1 peptides. In Reference 14, the extent of outgrowth was measured by placing purified cells on microwell dishes. Some neurons were labeled using a fluorescent dye and added to the existing cell mixture. In the work of Powell et al.,[14] neurite length of the labeled cells was measured using a microscope and the average neurite length and number were calculated.

As discussed previously, the use of CNTs as chemical biosensors relies on the interaction of various biological molecules with nanotubes. In yet another application of

CNTs, they may be used to record electrical activity of neurons. In this approach, it is necessary to bind the CNTs in close proximity of the neuron. In accomplishing this task, it is useful to identify peptide sequences that have selective affinity for CNTs. Wang et al.[15] have considered such peptide sequences. The location of binding peptide sequences were carried out using the phage display technique.[15] In this technique, the peptide is fused on the exterior of the bacteriophage and this was repeated with different peptide sequences. The bacteriophages were then suspended in detergent solution in the presence of nanotubes. This mixture was incubated for about an hour at room temperature. High speed spinning was used to facilitate the removal of the unbound phage particles. Incubation followed by centrifugation promoted the elution of the bound phage particles. The binding phage concentration was given as a measure of the number of plaque forming units (PFU). The larger the PFU value, the stronger the binding. In order to establish a direct proof of binding, the phage clones were amplified and coated onto microspheres using an antibody. The microspheres were then incubated with SWNTs and analyzed using a scanning electron microscope. Specific peptide sequences were determined by conducting phage display experiments on single-crystal graphite. One such sequence established was WPHHPHAAHTIR. The binding strengths of these peptides were tested by introducing mutations in them. It was found that the mutated peptides were weakly bound to the nanotubes when compared to the original peptides. These results take on special significance in view of the ongoing international effects to use CNTs as nanoscale components in high-performance electronic systems.

Pantarotto et al.[16] have considered the possibility of using CNTs to realize peptide-nanotube-based vaccines. Fragment condensation and selective chemical ligation are techniques that have been employed successfully to bind peptides to CNTs. The fragment condensation method was used to bind a pentapeptide to CNTs while the latter method employed the use of a peptide isolated from the foot and mouth disease virus (FMDV). The FMDV peptide retained its antigenic characteristics after being bound to the CNT and this was confirmed using ELISA (Enzyme-Linked Immunosorbent Assay) and a surface plasmon resonance test. As discussed by Pantarotto et al.,[16] characterization of peptide bound CNTs was performed using TEM and NMR spectroscopy. In summary, these *in vivo* studies show that the FMDV peptide-CNT conjugate evokes an immune response and this strengthens the possibility of manufacturing peptide-nanotube based vaccines.

Carbon nanotubes, owing to their extremely small size and interesting electrical properties have potential applications as nanoscale components in instruments used to study nanostructures. Atomic force microscopy (AFM) is one such technique. AFM involves the use of probes to study the characteristic features of surfaces. Nanotubes possess a high aspect ratio and their use as the tip of the AFM makes for easier probing over the sample surface. Owing to their cylindrical geometry, nanotubes facilitate the imaging of narrow, deep structures. Sample damage is minimal since nanotubes have sizes comparable to molecules. CNT tips buckle if the force imparted exceeds a critical value. Also, the lateral resolutions offered by these tips are much higher when compared to the currently used Si tips.[17] The process of shortening of tubes results in the formation of tubes with open ends (confirmed by TEM). By functionalization of the tips of these nanotubes by various acidic and basic groups, CNTs may be used as probes to extract information with nanoscale resolution. By coupling carboxylic and amine groups at the tips of nanotubes, amide linkages can be formed at the tips of nanotubes. This was demonstrated by covalently linking biotin to these tubes by the formation of an amide bond. Also, the open ends of oxidized nanotubes possess carboxylic acid groups which can be coupled with amine groups. These modified tubes were then used to study the binding interactions between biotin and streptavidin. A control experiment performed with unmodified nanotubes did not indicate any biotin-streptavidin binding force. Such functionalization may also be extended[18] to SWNTs to further facilitate high-resolution imaging and mapping of surfaces.

7. COMBINING NANOELECTROMECHANICAL SYSTEMS (NEMS) AND MICROELECTROMECHANICAL SYSTEMS (MEMS)

Incorporating nanoscale devices onto microelectromechanical systems (MEMS) could enhance the performance of a variety of microelectromechanical systems. Williams et al.[19] discuss a method adopted to place a single carbon nanotube onto a MEMS structure. A combination of AFM (atomic force microscope) and SEM (scanning electron microscope) was used to isolate a single CNT from a network of tubes. The isolated CNT was then placed at a specific location in the MEMS device. The movement of the AFM tip and the surface-tip interactions were monitored closely with the aid of SEM imaging. MWNTs were manufactured by the arc discharge technique and the cartridges were made using copper electrodes. The copper cartridges were placed such that the tubes were perpendicular to the surface of the MEMS structure. A single CNT was isolated from the cartridge by bringing in close contact an AFM tip. The adherence of the nanotube to the AFM tip was attributed to the van der waals force of attraction. The isolated nanotube was then placed in the gap between the pointer and the reticle. The change in the shape of the nanotube indicated contact between the pointer and the CNT. Subsequent welding of the nanotubes onto the pointer by focusing an SEM electron beam resulted in the deposition of carbonaceous compounds at the junction. The other end of the tube was welded onto the reticle by the same technique. The application of voltage to the pointer indicated that current travels between the nanotube and the MEMS structure. The strength of the welded nanotube was tested by the application of strain at the nanotube ends. It was observed that the nanotubes flexed but remained intact indicating that they were firmly affixed at the ends. It was also shown (by the application of forces in the lateral direction) that the tensile strength of the tubes is greater than the strength of the welds. By increasing the amount of carbonaceous material at the junction and the duration of deposition, the strength of the welds can be enhanced. Some of the parameters that need to be estimated before the incorporation of nanotubes onto MEMS are their load carrying capacity and tensile strength. The optimization of these parameters would help achieve near-nano resolution in MEMS.

8. CONDUCTION IN MULTI-WALLED CARBON NANOTUBES

Multiwalled carbon nanotubes may be viewed as concentric shells of individual CNTs. The manufacture of MWNTs results in the formation of both metallic and semiconducting tubes. Moreover, these tubes tend to form a cluster, which is undesirable if these tubes are to be used as electronic devices. As demonstrated by Collins et al.,[20] the concentric shells of a MWNT can be separated from each other by current induced electrical breakdown. The current supplied must be high in order to overcome the strong carbon-carbon linkage. The induced current oxidizes and removes the outermost shell of a MWNT. A major portion of the current was seen distributed in the outermost shell as it is in direct contact with the external environment. As a result, the inner shells remain protected. The proof of shell removal was demonstrated both electrically and by the use of analytical techniques like SEM and AFM. The same technique can be used in the separation of semiconducting SWNTs from a mixture of metallic and semiconducting SWNTs. The breakdown observed in a mixture of SWNTs is not uniform as all the individual tubes are exposed to the external environment.[20] The contribution to conductance from the outermost shell can be calculated by studying I-V relations for a MWNT with n shells and n-1 shells. The MWNT studied by Collins consisted of a semiconducting outer shell and a metallic inner core. The determination of the contribution of the inner metallic portion of the nanotube to the total conductance is facilitated by removing the outer shell. In Reference 20, the conductance of a MWNT was studied. It was estimated that at room temperature, the semiconducting outer shell and the metallic core contributed to the conductance of the nanotube while at temperatures around 90 degrees kelvin, the contribution from the core was negligible as the metallic component was completely frozen.[21] In order to improve the electrical properties of SWNTs, deposition of metals was carried out using electron beam evaporation and the metal-tube interactions were studied. Among the metals that were studied are Ti, Ni, Al and Fe. It was established that Ti and Ni formed uniform layers on nanotube surfaces whereas Al and Fe existed as discontinuous films. TEM and SEM images of metal-coated nanotubes were used to study metal-nanotube interactions as well as the distribution characteristics of metals on nanotube surfaces.[22]

Fluctuations in electronic properties of CNTs occur mainly due to the interaction between nanotubes and the surfaces of substrates. These interactions occur primarily through van der Waals forces and have a substantial effect on the geometric properties of nanotubes. The effects of these forces on nanotube surfaces and the binding energies between the nanotubes and substrates have been established using AFM and molecular mechanics simulations. It was further established that the nanotubes undergo both axial and radial deformations depending on the nanotube diameter and the number of shells in the MWNTs. The deformations observed in large diameter nanotubes were greater than those observed in small ones. By increasing the number of inner shells, the binding energy was lowered and the extent of deformation observed was much less.[23]

9. RECENT DEVELOPMENTS IN CARBON NANOTECHNOLOGY

The small scale production of CNTs hardly poses any threat to the general public due to their limited exposure. Researchers have, however, investigated the question of whether an increase in the production rates of nanotubes could prove harmful in the longer run. Experiments have been carried out on mice by exposing them to SWNTs and carbon black.[24] Results obtained indicated that carbon black caused minimal damage but nanotubes, even in minute concentrations caused granuloma formation. Carbon particles, from both tubes and carbon black were detected in the air sacs of lungs (alveoli). Other nanoparticles, like the ones made from PTFE are also considered toxic. These particles (due to their extremely small sizes) cannot be removed easily from the body by macrophages. A drastic reduction in the size of these particles could alter them chemically. It is anticipated that additional studies of the toxicity of these nanoparticles as well as the toxicity of CNTs will be forthcoming.

By a slight modification in the manufacturing process of CNTs, nanotubes as long as 4 mm have been formed. This has been demonstrated by Jie Liu and his team at the Duke University as well as by Saveliev et al. who use a methane oxygen flame for CNT synthesis.[25] In the standard chemical vapor deposition (CVD) process, the furnace is warmed from room temperature to about 900 degree centigrade resulting in the formation of nanotubes of about 20 micrometers in length.[25] Preheating the furnace to 900 degrees centigrade before placing the catalysts resulted in lesser aggregation of catalysts and longer tube formation. This was due to reduction in time for catalyst heating from 10 minutes to a few seconds. The modified technique is expected to facilitate alignment of nanotubes in a two-dimensional grid as well as the potential applications of these nanotubes are as nanoscale components in both biosensors and in nanoscale transistors.[25]

One of the major obstacles experienced by nanotube researchers was differentiating metallic CNTs from semiconducting CNTs. An electrical technique to separate the two types of tubes has been suggested and implemented successfully.[26] Since metallic nanotubes may be dimensionally very similar to semiconducting nanotubes, the differences in their electrical properties were exploited to sort them from a mixture. When placed in a direct current electric field, both types of tubes formed dipoles with positive and negative charges accumulating at opposite ends. But when placed under the influence of an alternating field, the rate of electron motility was much faster in metallic nanotubes than in semiconducting nanotubes. This resulted in quicker polarization and movement of metallic nanotubes (stronger dipoles) towards the electrode. This phenomenon was used to sort CNTs. The sorting technique has proved successful with minute quantities of nanotube mixtures and needs to be scaled up for processes that use larger nanotube volumes. Recently, a team of researchers at the Rice University led by Smalley[27] has discovered fluorescence effects in CNTs. As is well known, a principal characteristic of fluorescence is that the light emitted by an object being illuminated has a wavelength different from that of the incident beam. Moreover, it was observed that the wavelength of emitted light depends on the diameter of the CNT. This remarkable property could be combined with the biosensing capabilities of CNTs to detect and target specific cells of the body.

As is well known, one of the well-studied classes of quantum dots (QDs) is that of fluorescent semiconducting nanocrystals. These quantum dots have been used primarily in labeling and imaging of cells.[36] Coupling of these structures with MWNTs has been successfully carried out by Ravindran et al. and the complete procedure is dealt with in detail in their recent paper.[37] As reported, oxidation of MWNTs (under controlled conditions) in the presence of concentrated nitric acid results in the production of hydrophilic carboxylic acid groups at ends of the CNT. This was followed by the introduction of amine groups on ZnS-coated CdSe QDs with the aid of AET (2-aminoethane thiol hydrochloride). The ZnS coating shields the inner core and helps increase the quantum yield of the dots. MWNT-QD coupling was then carried out in the presence of EDC [1-ethyl-3-(3-dimethylaminopropyl) carbodiimide] through the formation of a sulfo-succinimidyl intermediate that serves as a cross-linking agent between these QDs and CNTs. Moreover, it was observed that the binding of QDs to the ends of the CNTs did not produce observable changes in the electronic properties of the CNT. The resulting CNT-QD conjugates are prototypes of the types of nanostructures that would convert and couple optical signals to electrical signals on the nanoscale. Clearly, such structures are likely to find applications in bioengineering.

10. ADDITIONAL APPLICATIONS OF CARBON NANOTUBES

The use of CNTs as artificial muscles represent yet another possible, if not immediate application.[28] These CNTs have abilities to convert electrical energy into motion and can generate and withstand extremely high stresses. The use of CNTs as high energy storage devices is another likely possibility that could be realized in the years to come. CNTs have the extraordinary ability to adsorb hydrogen onto their surfaces and adsorbed hydrogen is easier to pack densely than compressed hydrogen. Hydrogen may also be packed inside CNTs. Potential uses would be in the field of fuel cell technology where nanotubes are viewed as hydrogen storage devices.[29]

The chemical sensing capabilities of CNTs are being exploited in the detection of chemicals that are viewed as harmful pollutants. These tubes have extremely high surface areas that help in easy adsorption of chemicals. Carbon nanothermometers employ the use of liquid gallium inside nanotubes. The electrical properties of these nanotubes are being researched and the use of CNTs as nanotweezers, infrared sensors and chemical actuators have been proposed.[3]

11. CONCLUSION

CNTs possess extraordinary mechanical, optical and electrical properties, which make them potentially useful for applications in fields ranging from electronics to biotechnology. Standard manufacturing methods result in the formation of both metallic and semiconducting nanotubes. A major concern has been the isolation of the metallic ones from their semiconducting counterparts. The recent discovery of a means of sorting nanotubes using a novel electrical technique has contributed to the solution of this problem. Summarizing the biological applications of CNTs has been the main focus of this chapter. The use of these nanotubes as sensors has drawn the attention of various researchers and various methods to functionalize and solubilize these nanotubes have been proposed. The remarkable mechanical properties of these nanostructures have been exploited in the use of these tubes as tips for analytical techniques such as atomic force microscopy and scanning probe microscopy. Moreover, these nanotubes may be used to analyze organic samples without damaging them as they buckle if the force imparted exceeds the critical value. Functionalization of tips with carboxyl and amine groups would help in high precision mapping and imaging of sample surfaces. In the field of neuroscience, they portend extensive use in the recording of electrical activity in neurons and as substrates for neuronal growth. The recent discovery of CNTs with DNA sensing capabilities has been a major step in the use of CNTs as biosensors. In order to enhance the biosensing properties of CNTs, the tube surfaces have been successfully functionalized using lipids and surfactants.[30] The transition from MEMS to NEMS represents a major challenge. Just as with nanotechnology in general, the technological revolution based on CNTs is in its infancy. If present trends continue, the ever increasing potential for using CNTs in bioengineering will flourish in the decades to come.

12. ACKNOWLEDGEMENTS

We thank Omprakash Muppirala and Narenkumar Pandian for their help in preparing Figure 3 and Figure 4.

13. REFERENCES

1. C. R. Martin, and P. Kohli, The emerging field of nanotube biotechnology, *Nature Reviews Drug Discovery* **2**(1), 29-37 (2003).
2. M. Dresselhaus, G. Dresselhaus, P. Eklund, and R. Saito, Carbon Nanotubes, *Physics World,* Jan 1998.
3. Juthika Basak, Deepanjan Mitra, and Shashank Sinha, University of California, LA, Carbon Nanotubes (CNTs): The Next Generation Sensors.
4. R. J. Chen, Y. Zhang, D. Wang, and H. Dai, Noncovalent sidewall functionalization of single walled carbon nanotubes for protein immobilization, *J. Am Chem Soc.* **123**(16),3838-3839 (2001).
5. R. J. Chen, S. Bangsaruntip, K. A. Drouvalakis, N. W. Kam, M. Shim, Y. Li, W. Kim, P. J. Utz, and H. Dai, Noncovalent functionalization of carbon nanotubes for highly specific electronic biosensors, *Proceedings of the National Academy of Sciences* **100**(9), 4984-4989 (2003).
6. Moonsub Shim, Nadine Wong Shi Kam, Robert J. Chen, Yiming Li, and Hongjie Dai, Functionalization of Carbon Nanotubes for Biocompatibility and Biomolecular Recognition, *Nano Letters* **2**(4), 285-288 (2002).
7. B. R. Azamian, J. J. Davis, K. S. Coleman, C. B. Bagshaw, and M. L. Green, Bioelectrochemical single-walled carbon nanotubes, *J. Am. Chem. Soc.* **124**(43), 12664-12665 (2002).
8. M. P. Mattson, R. C. Haddon, and A. M. Rao, Molecular functionalization of carbon nanotubes and use as substrates for neuronal growth, *J. Molecular Neurosci.* **14**(3), 175-182 (2000).
9. K. A. Williams, P. T. Veenhuizen, B. G. de la Torre, R. Eritja, and C. Dekker, Carbon nanotubes with DNA recognition, *Nature* **420**, 761-763 (2002).
10. Michael J. O'Connell, Peter Boul, Lars M. Ericson, Chad Huffman, Yuhuang Wang, Erik Haroz, Cynthia Kuper, Jim Tour, Kevin D. Ausman, and Richard E. Smalley, Reversible water solubilization of single walled carbon nanotubes by polymer wrapping, *Chemical Physics Letters* **342**, 265-271 (2001).
11. E. T. Mickelson, C. B. Huffman, A.G.Rinzler, R. E. Smalley, R. H. Hauge, and J. L. Margrave, Fluorination of single-wall carbon nanotubes, *Chemical Physics Letters* **296**, 188-194 (1998).
12. P. J. Boul, J. Liu, E. T. Mickelson, C. B. Huffman, L. M. Ericson, I. W. Chiang, K. A. Smith, D. T. Colbert, R. H. Hauge, J. L. Margrave, and R. E. Smalley, Reversible sidewall functionalization of buckytubes, *Chemical Physics Letters* **310**, 367-372 (1999).
13. Sugary ways to make Nanotubes Dissolve, Chemical and Engineering News, July 9, (2002).
14. Sharon K. Powell, Jayashree Rao, Eva Roque, Motoyoshi Nomizu, Yuichiro Kuratomi, Yoshihiko Yamada, and Hynda K. Kleinman, Neural Cell Response to multiple Novel Sites on Laminin-1, *Journal of Neuroscience Research* **61**, 302-312 (2000).
15. Siqun Wang, Elen S.Humphreys, Sung-Yoon Chung, Daniel F.Deluco, Steven R. Lustig, Hong Wang, Kimberly N. Parker, Nancy W. Rizzo, Shekar Subramoney, Yet-Ming Chiang, and Anand Jagota, Peptides with selective affinity for carbon nanotubes, *Nature Materials* **2**, 196-200 (2003).
16. Davide Pantarotto, Charalambos D. Partidos, Roland Graff, Johan Hoebeke, Jean-Paul Briand, Maurizio Prato, and Alberto Bianco, Synthesis, Structural characterization, and immunological properties of carbon nanotubes functionalized with peptides, *J. Am. Chem. Soc.* **125**, 6160-6164 (2003).
17. Stanislaus S. Wong, Adam T. Woolley, Teri Wang Odom, Jin-Lin Huang, Philip Kim, Dimitri V. Vezenov, and Charles M. Lieber, Single-walled carbon nanotube probes for high-resolution nanostructure imaging, *Applied Physics Letters* **73**(23), 3465-67 (1998).
18. S. S. Wong, E. Joselevich, A. T. Woolley, C. L. Cheung, and C. M. Lieber, Covalently functionalized nanotubes as nanometre-sized probes in chemistry and biology, *Nature* **394**(6688), 52-54 (1998).
19. P. A. Williams, S. J. Papadakis, M. R. Falvo, A. M. Patel, M. Sinclair, A. Seeger, A. Helser, R. M Taylor, S. Washburn, and R. Superfine, Controlled placement of an individual carbon nanotube onto a microelectromechanical structure, *Applied Physics Letters* **80**(14), 2574-2576 (2002).
20. P. G. Collins, M. S. Arnold, and P. Avouris, Engineering Carbon Nanotubes and Nanotube Circuits Using Electrical Breakdown, *Science* **292**, 706-709 (2001).
21. P. G. Collins, and P. Avouris, Multishell conduction in multiwalled carbon nanotubes, *Appl. Phys.* A**74**, 1-4 (2002).

22. Y. Zhang, Nathan W. Franklin, Robert J. Chen, and Hongjie Dai, Metal coatings on suspended carbon nanotubes and its implication to metal-tube interaction, *Chemical Physics Letters* **331**, 35-41 (2000).
23. Tobias Hertel, Robert E. Walkup, and Phaedon Avouris, Deformation of carbon nanotubes by surface van der Waals forces, *Physical Review* B **58** (20), 13870-13873 (1998).
24. R. F. Service, Nanomaterials show signs of toxicity, *Science,* **300**(5617), 243 (2003).
25. A.V. Saveliev, W. Merchan-Merchan, and L.A.Kennedy, "Metal catalyzed synthesis of carbon nanostructures in opposed flow methane oxygen flame", Combustion Flame **135**, 27-33 (2003).
26. R. F. Service, Sorting technique may boost nanotube research, *Science* **300**, 2018 (2003).
27. http://www.rice.edu/projects/reno/Newsrel/2003/20020729_fluoresce.shtml.
28. R. H. Baughman, C. Cui, A. A. Zakhidov, Z. Iqbal, J. N. Barisci, G. M. Spinks, G. G. Wallace, A. Mazzoldi, D. De Rossi, A. G. Rinzler, O. Jaschinski, S. Roth, and M. Kertesz, Carbon nanotube actuators, *Science* **284**(5418), 1340-1344 (1999).
29. http://www.post-gazette.com/healthscience/19991011nanotubes1.asp.
30. Cyrille Richard, Fabrice Balavoine, Patrick Schultz, Thomas W. Ebbesen, and Charles Mioskowski, Supramolecular self-assembly of lipid derivatives on carbon nanotubes, *Science* **30**, 775 -778 (2003).
31. T. M. Chang, and S. Prakash, Procedures for microencapsulation of enzymes, cells and genetically engineered microorganisms, *Mol. Biotechnol.* **17**(3), 249-260 (2001).
32. C. R Martin, Nanomaterials - a membrane-based synthetic approach, *Science* **266**, 1961-1966 (1994).
33. D. T. Mitchell, S. B. Lee, L. Trofin, N. Li, T. K. Nevanen, H. Soderlund, and C. R. Martin, Smart nanotubes for bioseparations and biocatalysis, *J. Am. Chem. Soc.* **124**, 11864-11865 (2002).
34. C. Kneuer, M. Sameti, U. Bakowsky, T. Schiestel, H. Schirra, H. Schmidt, and C. M. Lehr, A nonviral DNA delivery system based on surface modified silica-nanoparticles can efficiently transfect cells in vitro, *Bioconj. Chem.* **11**, 926-932 (2000).
35. Jian Ping Lu and Jie Han, Carbon nanotubes and nanotube-based nano devices, in *Quantum-Based Electronic Devices and Systems*, edited by Mitra Dutta and Michael A. Stroscio, (World Scientific Publishing Co., Singapore, 1998), pages 101-123.
36. Gao Xiaohu, Warren C. W. Chan, and Shuming Nie, Quantum-dot nanocrystals for ultrasensitive biological labeling and multicolor optical encoding, *Journal of Biomedical Optics* **7**(4), 532-537 (2002).
37. S. Ravindran, S. Chaudhary, B. Colburn, M. Ozkan, and C. S. Ozkan, "Covalent coupling of quantum dots to multiwalled carbon nanotubes for electronic device applications, *NanoLetters* **3**, 447-453 (2003).

NANOPHYSICAL PROPERTIES OF LIVING CELLS

The Cytoskeleton

Gregory Yourek, Adel Al-Hadlaq, Rupal Patel, Susan McCormick, Gwendolen C. Reilly, Jeremy J. Mao*

1. INTRODUCTION

This chapter concerns the nanostructural and nanomechanical characteristics of the cell membrane of bone marrow derived mesenchymal stem cells (MSCs) and how cell membrane nanocharacteristics can affect cytoskeletal changes in response to flow-induced shear stresses. The present approaches are motivated by recent advances in both the fields of cellular micromechanics and cell-based tissue engineering. For example, computational models and experimental evidence have converged to support the notion that tissue-borne mechanical stresses outside the cell can be transmitted via the cytoskeleton to the cell nucleus.[1, 2] We have recently characterized the nanostructural and nanomechanical properties of marrow-derived mesenchymal stem cells.[3] Further evidence has suggested that exogenous mechanical stresses can upregulate and downregulate an increasing amount of mechanosensitive genes.[4-8] On the tissue-engineering front, adult MSCs have been differentiated into chondrogenic and osteogenic lineages and have been encapsulated into two stratified layers in hydrogel polymers, leading to tissue-engineered neogenesis of human-shaped articular condyle.[9] Despite these advances, a fundamental aspect of mesenchymal stem cells has been understudied – their biophysical characteristics in response to mechanical stresses. Because MSCs give rise to all skeletal tissues such as bone, cartilage, skeletal muscle, tendons, and ligaments, mechanical

* Gregory Yourek, Tissue Engineering Laboratory, Depts of Bioengineering and Orthodontics, Univ of IL at Chicago, Chicago, IL 60612. Adel Al-Hadlaq, Tissue Engineering Laboratory, Depts of Anatomy and Physiology and Orthodontics, Univ of IL at Chicago, Chicago, IL 60612. Rupal Patel, Tissue Engineering Laboratory, Depts of Dentistry and Orthodontics, Univ of IL at Chicago, Chicago, IL 60612. Susan M. McCormick, Depts of Bioengineering, Univ of IL at Chicago, Chicago, IL 60612. Gwendolen C. Reilly, Tissue Engineering Laboratory, Depts of Bioengineering and Orthodontics, Univ of IL at Chicago, Chicago, IL 60612. Jeremy J. Mao, Tissue Engineering Laboratory, Depts of Bioengineering and Orthodontics, Univ of IL at Chicago, Chicago, IL 60612.

modulation of MSCs is of paramount importance to their differentiation behavior. Hence, there is much interest in investigating the effects of the nanostructural and nanomechanical characteristics of MSCs on their differentiation and metabolism, using combined biological and engineering approaches. In this chapter we describe our current studies using stem cell culture and differentiation, atomic force microscopy, and application of shear stresses. We hope that the present combined biological and bioengineering approaches, will provide much needed insight into how adult stem cells are regulated by micromechanical stresses toward purposeful differentiation pathways.

2. PHYSIOLOGICAL ASPECTS

2.1. Human Bone Marrow Stromal Cells

Mesenchymal stem cells (MSCs), also known as bone marrow stromal cells (BMSCs), or mesenchymal progenitor cells, are defined as self-renewing multipotential cells with the capacity to differentiate into several distinct mesenchymal cell types.[10] MSCs represent a group of mesenchymal precursors that contribute to the regeneration of bone, cartilage, adipose, tendon, muscle, neural, and other connective tissues.[11-18] Suspensions of MSCs form a fibrous osteogenic tissue when incubated within diffusion chambers implanted *in vivo*.[19] This implies that the differentiating capacity of these cells is not dependent on the structural relationship of the cells *in situ* and that cells capable of differentiating in an osteogenic direction are present in marrow stroma.[19] When expanded *ex vivo*, MSCs have been shown to generate a bone-like tissue.[20-22] MSCs can be directed towards osteogenic differentiation *in vitro* when cultured in the presence of dexamethasone, β-glycerophosphate, and ascorbic acid.[23] There is much interest in the potential of using stem cells populations in the tissue engineering of bone and cartilage for the treatment of musculoskeletal trauma and disease. Adult mesenchymal stem cells offer certain advantages over embryonic stem cells for tissue engineering including their readiness and availability as they can be obtained from the same individual.[24] The studies described here were designed to investigate the biophysical and nanostructural characteristics of MSCs, in parallel with recent meritorious effort on biochemical characterization.[25-28]

2.2. Cytoskeleton

The cytoskeleton plays an important role in cell morphology, adhesion, growth, and signaling. The actin cytoskeleton consists of 3 components, actin filaments, intermediate filaments and microtubules. The "backbone" of the cytoskeleton is the actin filaments composed of F-actin. Actin filaments consist of repeating subunits of monomers arranged in a right-handed double helical structure. The widths of these filaments have a diameter of 5-9 nm.[29] Actin filaments are dynamic structures and can be elongated or shortened by polymerization or dissociation of monomers at the filaments ends[29], this creates a "treadmilling" effect, in which one end is polymerized while the other is depolymerized. Filaments can be bundled or cross-linked to each other by several actin binding proteins to create a network. Actin and intermediate filament systems have been described as "molecular guy wires" to mechanically stiffen the nucleus and hold it in place.[30] There is a direct link between integrins, the actin cytoskeleton, and the nucleus.[30]

The actin network plays a major role in the determination of the mechanical properties of living cells.[31, 32] Changes within the cytoskeleton of the cell allow the cell to migrate, divide, or maintain its shape.[33, 34] Actin polymerization and depolymerization are important actions in cell signaling and metabolism, as has been shown by subjecting cells to actin disrupting drugs such as cytochalasin D (cytD), latrunculin A (latA), or Jasplakinolide.[35-37] The effects of cytochalasins on cell elasticity has been studied by both a device designed to poke individual cells[38] and more quantitatively by AFM.[31] Rotsch and Radmacher[31] demonstrated the importance of the actin network for mechanical stability of living cells by showing that disaggregation of actin filaments always resulted in a decrease in the cell's average elastic modulus.

2.3. Mechanobiology

An increase in bone mass and strength across multiple species occurs as a result of mechanical loading, especially associated with cyclic mechanical stresses.[6, 8, 39-41] Conversely, bone loss results from bedrest, immobilization, and weightlessness.[42-46] It seems clear that both bone resorption and bone formation are mediated by mechanical signals. Among the multiple theories regarding which is the specific physical signal to which bone cells respond, strain-induced fluid flow has received the most abundant experimental support. Fluid flow occurs in the interstitial spaces around bone cells due to circulation and repetitive loading and unloading of bones during activity.[47] It is hypothesized that the sensing mechanism in mechanical load-induced bone remodeling is that the load induced fluid flow produces shear stress at bone cell membranes and these shear stresses induce cell signaling events which translate the mechanical signal to a cellular response.[48] Changes in interstitial fluid flow due to internal bone pressure changes influence bone remodeling whereas normal pressures serve to maintain a baseline level of flow and bone maintenance.[49]

2.4. Fluid Flow

Fluid flow is a mechanical stimulus likely experienced by all cells. A fluid flow signal experienced at a cell membrane partially transduced into a response by the cytoskeleton. Either transient, or more likely sustained, signals may lead to upregulation and/or downregulation of certain genes. At least two mechanisms are likely to be involved in the transmission of these mechanical signals from the cell membrane to the cell. These processes are termed "mechanotransduction". The first depends on second messengers released close to the membrane that then activate a chemical reaction cascade, leading to transcription factor activation and translocation into the nucleus. The second is that direct cytoskeletal displacement is transmitted to the nucleus and other cell communication points through physical coupling rather than chemical reaction cascades. It has been hypothesized that in mature osteoblasts, fluid flow creates a drag on the glycocalyx (a proteoglycan rich cell coat) and this drag causes the cytoskeleton to be stimulated and transmit a signal to the cell nucleus.[1, 2] Experimental data show that a proteoglycan rich glycocalyx is necessary for biological responses to fluid flow[50] and that the actin cytoskeleton is necessary for other biological responses to fluid flow.[5] However, although MSCs have been shown to also respond to a fluid flow stimulus,[51-54] it is not known how the MSC cytoskeleton differs from the mature osteoblast cytoskeleton or if the cytoskeleton can contribute to a fluid flow induced differentiation.

3. BASICS FOR STUDYING CONTRIBUTION OF CYTOSKELETON TO DIFFERENTIATION

3.1. Differentiating Factors

The pluripotency of MSCs has been demonstrated by exposing the cells to their normal growth medium, typically α-MEM (α-Minimum Essential Medium Eagle) or DMEM (Dulbecco's Modified Eagle Medium) plus 10-20% fetal bovine serum (FBS) and 1% antibiotics with supplementation by growth factors and other pharmacological agents. The virtually infinite expansion and differentiation of these cells into other cell types not as easily cultured indicate that MSCs will eventually play a major role in treatment for trauma, disease, or aging. MSCs have been successfully differentiated into adipocytes (fat cells), chondrocytes (cartilage cells), and osteoblasts (bone cells) by a number of laboratories under a wide variation of culture conditions, demonstrating the relative ease with which these cells can be utilized.

Stem cells have been differentiated towards an osteogenic lineage by supplementation with *(i)* dexamethasone, a member of the glucocorticoid family of steroids which may modulate Bone Morphogenetic Proteins (BMPs),[55] which are major inducers of osteogenesis, *(ii)* β-glycerophosphate to promote calcium phosphate deposition[9, 56-59] and *(iii)* ascorbic acid which plays an important role in the production of the collagenous bone extracellular matrix. Bone marrow stromal cells have been differentiated towards the chondrogenic lineage by supplementation with a number of different factors, the most potent of these being members of the transforming growth factor-β (TGF-β) family.[9, 23, 60, 61] For example, TGF-β1 is found in great supply in embryonic cartilage[62]. Finally, MSCs have been differentiated towards a adipogenic lineage by supplementation with dexamethasone, 1-methyl-3-isobutylxanthine (MIX), and indomethacin (both inhibitors of cAMP and reducers of the synthesis of lipogenic enzymes).[23, 60, 63]

3.2. Pharmacological Cytoskeleton Disrupting Agents

Pharmacological agents have recently become a tool for clinicians and researchers for disrupting the cytoskeleton of cells for both therapeutic and research reasons. These agents are derived from plant sources where they are used as defense mechanisms for otherwise defenseless organisms.[29] When consumed by an organism, these poisons attack one of the three cytoskeletal components (actin filaments, intermediate filaments, or microtubules) and cause great damage to the infected organism. When used in the laboratory for studies on the importance of cytoskeletal components in cellular events or in a clinic to combat certain types of cancer (for reviews see: 64, 65), they can be a useful mechanism for cell modification.

Cytochalasins are isolated from fungi and have an abundance of cytotoxic activites, including the disruption of actin filaments.[66] These compounds "cap" actin filaments by binding to the growing end of actin[67-72] and cleave actin filaments.[71, 73-75] There are agents which are specific to each of the three parts of the cytoskeleton. For example, phalloidin, cytochalasin, swinholide, and latrunculin have all been found to disrupt actin filaments, each acting through a specific mechanism.[29] Taxol (a chemotherapeutic), colchine, vinblastine, and nocodazole disrupt microtubules.[29]

3.3. Mechanical Stimulation of Cells

Parallel-plate flow chambers have been used to examine the effects of flow-induced shear stresses on the biochemical and morphological response of many types of cells. They provide a well-controlled mechanical stimulus of shear stress to a relatively homogeneous population of the cells of interest.[48-50, 76-82] Low level shear stress may have beneficial effects on cellular metabolism.[77] Fluid flow through parallel-plate flow chambers is laminar and has been well characterized[83] and since most of the apparatus is assembled of glass, polycarbonate, or rubber, there is no loss of medium due to permeation through the tubing or evaporation.[76]

Cone-and-plate viscometers have also been used to study the effects of shear stress on a monolayer of cells;[84-87] however, these systems have a smaller cell-to-volume ratio, do not permit continuous sampling of the cell culture medium and have significant culture medium evaporation, requiring continuous infusion of fresh medium.[76]

Recently, an apparatus known as Flexercell has been developed by Flexcell (Hillsborough, NC) that consists of six-well culture dishes with flexible silicone rubber bottoms which uses positive pressure to compress samples with a piston. Simmons et al.[88] used this system to study the effects of equibiaxial cyclic strain on hMSCs cultured in osteogenic media. They found a decrease in proliferation and an increase in matrix mineralization over unstrained cells. Flexcell has also been used by a group in the Netherlands[89] to show that the stage of differentiation that a osteogenic cell is at affects its response to stretch *in vitro* by noting differences in proliferation and apoptosis of a differentiating human fetal osteoblast cell line.

4. ATOMIC FORCE MICROCOPY (AFM)

4.1. Operation Principles

The fundamental principle of AFM imaging is to "sense" the sample surface with a probe. This "sensing" or probing of the sample surface is accomplished in a precisely controlled manner by the movement of a piezoelectric scanner, which has the sample mounted on top, against a micro-fabricated sharpened tip mounted at the end of a cantilever. The directional movement of the cantilever and the scanning tip is a function of the sample's surface topographic features including peaks and valleys, causing a deflection of the scanning tip as it moves across the sample surface. A laser diode tracks the cantilever movement by a laser beam focused on the cantilever surface that is opposite, but preferably over, the scanning tip. Therefore, each vertical and lateral deflection of the cantilever from the sample surface is coupled with a directional reflection of the laser beam. Most modern AFMs are equipped with a four-segment photodiode detector that registers the vertical and lateral reflections of the laser beam off the cantilever surface as a function of the laser intensity difference between different segments. These input photo signals are then amplified and transformed into AFM surface images through a digital control system.

Generally, AFM scanning results in two types of images: height image and force image. Whereas height images represent the sub-micron digital replication of the sample surface topographic characteristics, force images are a function of the elastic responses of the sample surface to nanoindentational forces applied on the surface by the scanning tip. Height and force images, individually or combined, provide fundamental characteristics of biological structures including cells and their surrounding matrices, thus offering great promise for elucidation of rich structural and functional details at unparalleled levels of resolution.

Another distinctive feature of the AFM is substantial minimization of sample preparation in comparison with other imaging modalities. Compared with electron microscopy, three dimensional AFM images are obtainable without expensive and time-consuming sample preparation, and yet yield more detailed representation than the two dimensional profiles available from cross-sectioned samples. AFM imaging eliminates the need for vacuum that is required by most electron microscopes to image samples, and yet has the potential to image samples in ambient air or liquid. The AFM's ability to view and scan samples in their quasi-native environment has resulted in the unsurpassed advantage of real-time imaging of live biological samples. In addition, compared to SEM that requires sample coating, AFM provides extraordinary topographic contrast with direct height measurements and unobscured view of surface features. Moreover, the extended ability of the AFM to measure the viscoelastic behavior of samples, where applicable, has broadened the versatility and implications of the AFM in various scientific fields.

Despite its advantages, AFM has several practical limitations. For example, because the depth of field in AFM is limited by the travel distance of the scanning tube (about 5.3 μm in the current model), AFM is best suited for imaging relatively soft samples. Depth of field is also limited by the relationship between the scanning tip geometry and the sample's surface topography. If the tip is too thick to reach into a groove or recess on the surface, a true image cannot be produced and thus the depth of field is reduced. The tip geometry relative to the surface features presents another challenge. Sheer walls and undercuts of the sample's surface may represent difficulties during AFM scanning due to the nature of the surface topography and tip geometry. Thus, it is critical to select appropriate AFM scanning tips for various biological samples.

4.2. Preparation of Cells for AFM Imaging

In general, the AFM can be used to scan and image any cell type provided that the cells can adhere or attach to a substrate to be mounted on the AFM scanner. Cells floating in suspension or lacking substrate support to withstand the interactions with the scanning tip are not suitable for AFM imaging.[90] The amount of the glycocalyx halo surrounding the cells[29] can present another limiting factor in obtaining a high-resolution image of the cell surface. Similarly, rapid cellular events that exceed the time required to obtain an AFM image, such as rapid cell growth, division, or movement, can result in low quality imaging. Typically, obtaining an AFM topographic image with acceptable resolving power for an area of about 10 μm^2 requires at least 2-3 minutes, and thus only slower cellular events are considered suitable candidates for the real-time AFM imaging.

A widely used method to prepare living cells for AFM imaging is to culture them on glass coverslips, usually treated with a material that facilitates cellular attachment such as the positively charged poly-D-lysine.[3] Also, trapping cells in millipore filters with pore size comparable to the dimensions of the cell can be used to image rounded living cells and to overcome the attachment requirement.[91] Another established methodology involves the use of AFM in conjunction with micropipette aspiration for cellular immobilization.[92]

4.3. Imaging Conditions

The driving motive to image samples in fluid is to study biological samples under normal physiologic conditions by minimizing surface interactive forces on soft samples. AFM imaging procedures using fluid are similar to those under air, but with several notable differences. Fluid imaging with the AFM minimizes the attractive surface tension forces between the scanning tip and the sample surface in the contact mode imaging, and thus also minimizes the scanning forces exerted by the cantilever and the tip deflection on the sample surface. Fluid imaging has shown its applicability especially for larger-scaled biological structures such as cells and extracellular matrices.

The imaging under fluid generally requires an additional piece of hardware such as a fluid cell to contain the sample and the imaging medium. The fluid cell is a small glass assembly that houses the liquid phase. The glass surface provides a flat beveled interface to avoid distortion of the AFM laser beam from fluid movement. The sample is mounted on a metal disk that fits magnetically on top of the AFM piezoscanner in a similar fashion to imaging in ambient air. The AFM fluid cell fits over the sample/disk assembly and holds the cantilever in a position above the sample. Fluid imaging can be performed by means of fluid cell with or without O-ring seal. Certain criteria are worth considering when choosing a liquid medium to image living cells with the AFM. For example, the liquid should not be too viscous, so to avoid potential interferences with tip/sample interactions and consequent image distortion. The liquid should also be clear of particles that can build up on the sample surface or the tip and affect their proper interaction. When imaging living cells, chemical and physical characteristics of the imaging liquid should resemble, as close as possible, physiologic conditions of the native extracellular fluids. Phosphate saline buffer (PBS) is commonly available and relatively inexpensive, and thus has broad acceptance among AFM users as a suitable imaging liquid-medium. However, when PBS dries, salt crystals form, making accurate scanning difficult. Regular serum-free culture medium, Dulbecco's Modified Eagle's culture Medium (DMEM) for example, or a mixture of DMEM and PBS can also suffice the imaging requirement. Whichever liquid medium is used, the pH of the solution should be maintained constant and within the normal physiologic range by frequent exchange or continuous flow of medium in case of extended imaging sessions.

4.4. AFM Probe Selection

The final AFM image is virtually a composite or "convolution" between the geometric properties of the tip and the sample surface.[93] Undoubtedly, one of the most important controllable parameter for AFM imaging, either in ambient air or fluid, is proper selection of scanning tips. The AFM scanning "tip" typically consists of a sharp microfabricated barb or spike mounted at the end of a "cantilever" to form a unified "probe". Although these terms are sometimes used interchangeably, only the microstructured component at the end of the cantilever is the part that actually indents the sample surface during the AFM scanning. A wide variety of AFM scanning tips are currently available, differing in geometry, material properties, and chemical composition.

Tips are broadly defined by their aspect ratio (length to width), opening angle and/or radius of curvature. Relatively rough sample surfaces, in the range of micrometers, should be scanned using high-aspect-ratio tips, which combine small opening angles with a long tip. However, low-aspect-ratio tips, corresponding with high opening angles, are more suitable for scanning relatively flat specimens.[94, 95] A special type of low-aspect-ratio tips exist where the very end is shaped into a high-aspect-ratio peak with overall low-aspect-ratio configuration of the tip. These tips are referred to as "sharpened tips" and ensure greater scan depth with improved resolution when scanning relatively flat samples. Radius of curvature of the AFM tip reflects the nanometric sharpness of the tip's peak. Typical radius of curvature of sharpened tips is less than 20 nm, while that of unsharpened tips ranges from 20-50 nm. Oxide sharpened silicone nitride tips are widely utilized in the AFM scanning of living cells and various other biological structures[3, 96, 97] due to their high versatility and ability to combine high resolving power with physical tolerance on soft sample surface.

The tip-cantilever assembly is most commonly made of crystal silicone or silicone nitride, which are both suitable for microfabrication due to their stiffness and wear resistance. Silicone nitride tips are more suitable for contact mode imaging due to their flexibility and "forgiveness" on the sample surface compared to the stiffer crystal silicone probes. Another distinctive characteristic of silicone nitride probes is the greater tendency of the silicone nitride tips to be trapped by the surface tension attractive forces during interactions with the sample surface than the crystal silicon probes. Such forces, although micro- or nano-scale in nature, might be strong enough to deform the surface of soft samples. Therefore, considerable attention to the selection of the scanning tip should be taken, especially when imaging delicate samples. By contrast, when scanning harder samples or using tapping mode AFM, stiffer crystal silicone probes are likely more appropriate. However, increased brittleness of the crystal silicone tips due to their greater stiffness mandates considerable care during tip handling and preparation for the scanning session.

Several recent efforts are directed toward substituting the silicone and the silicone nitride with more characterized materials for the fabrication of enhanced AFM probes. Carbon nanotubes[98-101] are gaining rising popularity to be the backbone structural material for the second generation of AFM probes. Additional advantages offered by the carbon nanotubes include their well-characterized structure, mechanical robustness, and unique chemical properties that allow well-defined surface modification without jeopardizing the AFM scanning resolution. For example, utilizing this last feature of feasibility of carbon nanotubes' surface modification under high controllability, many aspects of structural and thermodynamic properties of protein-protein and protein-nucleic acid complexes interactions have been revealed.[102, 103] Moreover, besides their use for AFM imaging, a recent attempt introduced a novel chromosomal dissection method using carbon nanotube probes under direct AFM imaging.[104]

4.4. Scanning Modes

Based on the nature of the interactions between the scanning tip and sample surface, AFM scanning can be in contact mode, tapping mode, or error-signal mode. Mode selection largely depends on the nature of the sample and the desirable images to be obtained by the AFM.

In contact mode, the scanning tip makes a direct contact with the surface of the sample throughout the scanning period with the topographic features of the sample's surface dictating the degree of cantilever reflection. However, since the amount of cantilever reflection is pre-adjusted through the control system to a certain value (the operating set-point), this imaging mode is also known as "constant-force mode". In addition to the fact that this mode is the original AFM imaging mode and can be readily accomplished for a large variety of samples, the most advantageous feature of contact-mode AFM is its ability to perform the scanning under both air and fluid conditions. Fluid imaging with contact mode AFM is necessary for imaging living cells in an appropriate fluidic medium so that cell viability can be maintained in quasi-physiologic conditions.[93] Contact-mode AFM imaging in fluid is also beneficial in eliminating the capillary action forces, adding to the precision of the scanning force.[94] On the other hand, a disadvantage associated with contact-mode AFM scanning is the lateral frictional forces due to direct contact between the tip and the sample surface. In the case of soft samples, such as the cell surface, this direct contact may damage the structures of the imaged surface.[105] In contact-mode AFM, the deflection signals from the cantilever in the z direction are plotted against x and y to produce informative height images that reflect the nanometric height variations of the scanned surface.

Also known as "oscillating mode" or "non-contact mode", in tapping mode the scanning tip is literally bouncing up and down or "tapping" as it travels across the sample surface. The driving principle behind implementation of tapping mode is to eliminate the lateral shear forces associated with imaging in contact mode.[105] However, similar to contact-mode AFM, the vertical movement of the cantilever is maintained at constant oscillation amplitude throughout the scanning period. Similarly, tapping-mode AFM can be performed in air or fluid environment. The imaging of living cells in aqueous environment using tapping mode may actually have the potential of minimizing the frictional forces between the tip and the relatively soft surface of the cell membrane and thus can reduce the accumulation of membrane structures on the scanning tip that might hinder the scanning resolution.[90] However, for AFM topographic imaging, the use of the tapping scanning mode may result in less accurate duplication of surface topographic features since the deflection of the scanning tip is predetermined and is not dependent on the height variations of the scanned surface.

Error signal mode is the scanning mode of choice to get the most accurate reproduction of the surface topographic features, especially when scanning relatively rough and rigid sample surfaces such as that of cells or bacteria.[94] Error-signal mode or "deflection mode" derives its name from the fact that the operating set-point (the scanning force) is reduced to the lowest value and therefore the input signals that are translated into the resulting surface image are virtually registered by the amount of deviation, or "error", of the scanning tip and the cantilever in response to the surface features. This cantilever deflection as a result of encountering height variations on the sample surface with minimal "interference" or control from the feedback loop system, translates into scanning forces that are largely dictated by the sample surface roughness and hence more accurate reproduction of the surface topographic features are produced. The resulting topographic image is then literally a surface force map where high spots on the surface are represented by areas of high force on the image as a result of greater deflection of the cantilever and the similarly minimal cantilever deflection will be recorded as a region low force. Although more detailed topographic images can be obtained by utilizing the error-signal mode, one must be cautious when scanning living cells of any cell membrane disintegration as a result of high "poking" force in response to greater deflection from the scanning tip. A number of cell types and fine structural details of the cell membrane have been exposed by successfully employing the error-signal AFM scanning mode.[96, 106-108]

5. TOPOGRAPHIC IMAGING OF LIVING CELLS WITH AFM

To date, many cell types have been imaged successfully with AFM under quasi-physiologic conditions (for reviews: 93, 97, 105, 108-111). Topographic images are probably the most commonly acquired AFM images to analyze the sub-micron structural complexity of various biological surfaces, including cell membrane structures.[103, 112, 113] Topographic images represent the output data for the vertical deflection of the cantilever tip in response to encountering surface height variations, where the input signals are amplified and digitally translated into topographic images by the AFM control system. The extended ability to derive and analyze the surface roughness of the sample utilizing routine AFM height images has significantly supplemented the physical characterization process of the cellular membrane. Since early utilization of AFM for imaging living cells, the subsurface cytoskeletal structures have been observed and described in the nanometric-scale range.[114, 115] The cytoskeleton most readily resolved by the AFM is actin filaments.[116] The conjunction of the AFM with other imaging techniques has also confirmed the ability to study microtubules and intermediate filaments with the AFM.[117-120] Tightly adherent cells are stiffer than cells that are loosely attached,[118] suggesting a dynamic reorganization of the cytoskeletal elements induced by the cellular attachment to the substrate. Also, the portion of the plasma membrane overlying the nucleus is 10 times softer than the rest of the membrane in living fibroblasts.[117] Upon study of the three cytoskeletal elements with immunofluorescent dyes using confocal laser scanning microscopy, the elasticity of the cell membrane is related to the distribution of actin and intermediate filaments, but much less to microtubules,[117] Similar observations in two fibroblast cell lines confirmed the crucial importance of the actin filament network for the mechanical stability of living cells,[31] Whereas the disaggregation of actin filaments causes a decrease in the average elastic modulus of the cell membrane, induced disassembly of microtubules has little effect on cell membrane elasticity,[31] The relative contribution of the cytoskeleton to the elasticity of the cell membrane is also demonstrated in many other cell types, including epithelial cells,[106] cardiocytes,[121] astrocytes,[122] liver sinusoidal endothelial cells,[123] cancer cells and macrophages,[124] erythrocytes,[125] platelets,[126, 127] articular chondrocytes,[92] osteoblasts,[128] and a variety of other cell lines.[129-132]

Not only has the AFM provided a tool for observing the cytoskeletal elements with nanometric-scale resolution and analyzing their corresponding involvement in cell membrane elasticity, but it also has become a versatile instrument to assess the effect of different pharmacological agents on the cytoskeletal elements and cell membrane.[31, 133-135] In addition to the cytoskeleton, the AFM is gaining popularity for the biostructural analysis of microdomains and junctions that constitute the plasma membrane.[136-139] In particular, a promising approach is the constitutive amalgamation of selected molecules to the AFM tip to control the local interaction and recognition events between the AFM tip and certain molecules or receptors on the cell surface.[140-142] The implications of this approach extend beyond the structural analysis of local distribution of selected molecules to the specific mechanical characterization of molecules in the nanometric-scale range.[142] Indeed, the AFM's capability to characterize both the structural and mechanical properties of biological structures at the cellular and subcellular levels with nanometric resolution has provided an unprecedented tool for biological science.

6. FORCE IMAGING OF LIVING CELLS WITH AFM

The early use of the AFM to visualize living cells has been complemented by innovative incorporation of the AFM to derive the elastic properties of cells.[106, 143, 144] The elastic properties of cells can be determined by generating force images or "force-volume images" that represent the elastic behavior of the cell's surface or intracellular structures to well-controlled vertical probing forces applied to the surface by the AFM scanning tip. The cantilever deflection, as the tip probes areas with varying degrees of elasticity on the scanned surface, is reflected on the photodiode detector, where these signals are then transformed digitally into force image, representing the elastic response of various points within the scanned field of the sample surface in response to the introduced probing force. Force curves are generated automatically by selecting particular points on the force-volume image within each scanning field. Force curves typically consist of approaching and retraction phases, representing the arrival and departure motions of the AFM scanning tip over the sample surface. As the AFM probing tip approaches the sample surface, a number of interactive events take place before the tip makes an actual contact with the surface. Electrostatic forces between the scanning tip and the sample surface can be either repulsive or attractive, depending upon the electric charge of both the tip and sample surface. These repulsive or attractive forces are the first interactive forces between the material surface of the scanning tip and the outermost layer the sample surface. When scanning in fluid, as is the case when scanning living cells, the effect of tensile forces on the fluid surface should be taken into consideration as the liquid medium forms a thin layer between the approaching tip and the sample surface. As the tip moves closer to the sample surface, the Van Der Waals' attractive forces start to take effect between the molecules and atoms of the tip and the scanned surface. Finally, when the tip makes its way down to the sample surface, a nanoindentation on the surface is produced. The amount of the nanoindentation depends on several factors including the elastic behavior of the sample surface, nanoindentation force, and the type of the AFM scanning tip. Nanoindentation can be derived from force-displacement plots that are recorded by the AFM each time the tip approaches and retracts from the sample surface. The force displacement curve is simply a systematic translation for the deflection of the cantilever tip versus the z-direction displacement of the piezo-scanner as a result of the interaction between the AFM tip and the sample surface.

The Hertz model[145] is the most straightforward mathematical derivation for describing the elastic responses of an indented sample by the tip of the AFM.[146, 147] The Hertz model is described as:

$$E = \frac{3F(1-\nu^2)}{4\sqrt{R}\delta^{3/2}} \tag{1}$$

where E is the Young's modulus of elasticity, F is the applied nanomechanical load, ν is the Poisson's ratio, R is the radius of the curvature of the AFM tip, and δ is the amount of sample indentation. The amount of applied load and the indentation depth can be derived from force versus distance plots that are recorded by the AFM each time the tip approaches and retracts from the sample surface. The force versus distance curve is simply a systematic translation for the deflection of the cantilever tip versus the z-direction displacement of the piezo-scanner as a result of the interaction between the AFM tip and the sample surface.

The Poisson's ratio is a material property that describes the ratio of transverse deformation to the axial deformation of the material under axial loading.[148] Typical values for Poisson's ratio are in the range of 0.3 to 0.5.[97, 149] Thus, by substituting obtainable values in the Hertz model, it is possible to estimate the Young's modulus of elasticity of the studied sample. In the case of soft samples, such as living cells, one frequent limitation with mathematical fitting of force curves using the Hertz model is the difficulty in determining the tip-sample contact point due to (*i*) the softness of the cell surface and (*ii*) the lack of sharpness in curve deviation as the AFM tip contacts the sample surface as is the situation when the tip contacts a glass surface (Figure 1). Also, the legitimacy of the Hertz model for application to living cells is further depreciated by the highly anisotropic behavior of cellular systems[143]. Nonetheless, with a lack of more sophisticated analytical formulas to translate the AFM data into mechanical values, the Hertz model still presents a workable framework with valuable application to mechanical mapping of living cells, especially when a comparison of mechanical properties rather than absolute measurement of a mechanical value to be pursued.

Recently, the quadratic equation has proven to be a useful tool for modeling force-indentation curves.[32, 150] Applied forces, F, and resulting indentation depths up to 500 nm, δ, have been expressed by the quadratic equation:

$$F = a\delta^2 + b\delta \tag{2}$$

where a and b are the parameters expressing the nonlinearity and the initial stiffness of the force-indentation curve, respectively. Significant correlations ($r > 0.99$) have been found for the quadratically defined curves when compared with experimentally obtained force-indentation curves.[32, 150] As the thickness of a specimen increases, the parameter b decreases monotonically until a constant value is reached while an increase in the b parameter indicates an increase in mechanical stiffnes.[32] Both a and b parameters increase for sheared endothelial cells which may indicate a remodeling of the cytoskeletal structure since these cells exhibited thick stress fibers of F-actin bundles.[32] An elastic modulus, E_{FEM}, was determined using Finite Element Modeling (FEM). It's relationship to the quadratic equation variables is as follows:

$$E_{FEM} = \frac{b}{b'} \tag{3}$$

where b' is the linear coefficient. E_{FEM} increases in sheared endothelial cells.[32]

6.1. AFM in Cytoskeleton Imaging

AFM has made the structural study of the nanosized cytoskeleton elements possible. However, the study of the cytoskeleton comes at a cost. Low loading forces result in smooth versions of the cell's membrane, while the high forces uncover the true structure of the cell's cytoskeleton.[121] The high loading forces necessary to view the cytoskeletal

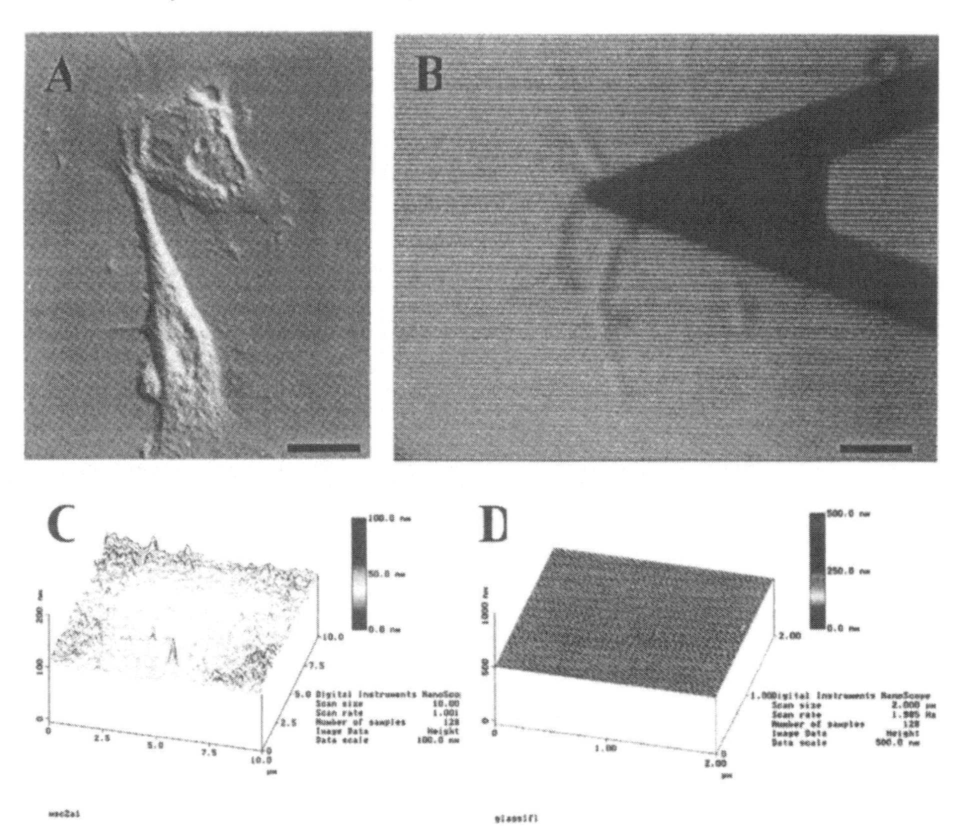

Figure 1. AFM imaging of living mesenchymal stem cells (MSCs) on glass. A. A single MSC during nanoindentation with V-shaped AFM cantilever. B. The height image representing the cellular extensions of two MSCs collected at a scan area of 20 μm² showing height variation within the scanning field. C. The height image of the glass substrate showing absence of height variation when cells are not present on the glass surface. D. The height image of the glass substrate with the fluid imaging medium present but without cells showing minimal height variation. (Bar = 20 μm)

structures cause a large lateral drag which damages the cell and make any preceding evaluation difficult. The importance of the actin cytoskeleton becomes readily apparent with its disruption with cytochalasin B. Chicken cardiomyocytes appear more rounded and a large decrease in the elastic modulus by about a factor of 3 occurred. This directly demonstrated that the elastic response of the chicken cardiomyocyte is due in a large part to the actin network.

Mechanical properties for multiple cell types have been reported including glial cells,[116] epithelial cells,[106, 151] cardiocytes,[121] myocytes,[152] platelets,[126] erythrocytes,[141, 153] macrophages,[154] endothelial cells,[32, 150, 155-157] fibroblasts,[31, 118, 130, 158, 159] osteoblasts,[128, 160, 161] chondrocytes,[92] hair cells,[162, 163] F9 carcinoma cell line,[164] and bone marrow-derived mesenchymal stem cells.[3] Mapping of mechanical properties of living cells can be viewed as an indicator of the cytoskeleton structure and function.[165] Multiple vital processes of the cell are dependent on the dynamics of the cytoskeleton, such as cell migration,[34, 166, 167] cell division,[168, 169] cell adhesion,[170, 171] cellular transport system,[172, 173] phagocytosis,[174, 175] and maintenance of overall stability and mechanical integrity of the cell.[176, 177] In addition to the cytoskeleton, numerous other molecules may also contribute to the overall mechanical map of the cell membrane and have an active role in determining the elastic properties of living cells.[140, 142, 178]

7. CELL NANOSTRUCTURAL CHANGES INDUCED BY DIFFERENTIATION OR MECHANICAL FORCES

7.1. Cell Differentiation

Wang et al.[179] used actin staining with palloidin[180] to demonstrate changes in the actin cytoskeleton upon differentiation of human MSCs with osteogenic (bone inducing) supplements (ascorbate, dexamethasone, β-glycerophosphate, and vitamin D_3). In addition to expressing markers specific to bone cells such as alkaline phosphatase, osteocalcin, and bone sialoprotein (BSP), production of collagen type I and bone sialoprotein and extracellular matrix mineralization, the differentiated cells displayed highly organized actin fibers alongside abundant bone sialoprotein complexes. In contrast to the latter, the undifferentiated MSCs had long actin filaments and sparse amounts of bone sialoprotein.

When the actin cytoskeleton of chick limb bud mesenchymal cells (CLBMCs) is disrupted chondrogenesis occurs. CLBMCs were isolated from stage 22-24 White Leghorn chick embryos and the cytoskeleton disrupted using cytochalasin D (see section 3.2).[181] The cells rounded up, lost their actin cables, and underwent chondrogenesis, as demonstrated by type II collagen and matrix production. These characteristics were amplified with increasing concentrations of cytochalasin as well as with introduction of the agent at earlier time points. However, the cells did not react this way with microtubule disrupting drugs, leading to the conclusion that actin filaments, rather than microtubules, control the shape of cultured limb bud cells and that these filaments play an important role in the early development of cartilage. Furthermore, the activation of the cell signaling molecule protein kinase C and the inhibition of the extracellular signal regulated kinase-1 has been shown to be involved in the regulation of chondrogenesis by actin cytoskeleton disruption by cytochalasin.[182, 183]

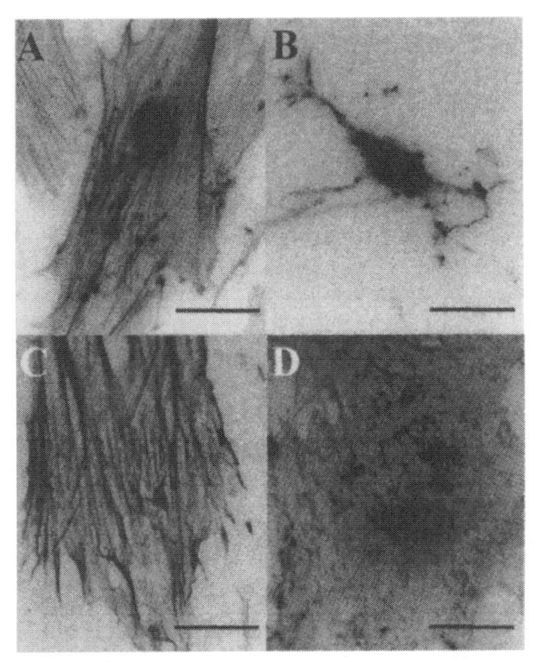

Figure 2. Human BMSCs in control (Con) or Osteogenic (OG) media 2 hr. after treatment with 0.5 μm Cytochalasin-D (cytD) or DMSO (Con). Actin network stained with rhodamine-labeled phalloidin (pictures have been converted to black and white so phalloidin shows dark in the images).
A. Con-Con **B.** Con-CytD **C.** OG-Con **D.** OG-CytD
Cells grown in osteogenic medium are more differentiated and have a more defined cytoskeleton which is less easily disrupted by cytD than that of undifferentiated hBMSCs. From this we can predict differences in nanomechanical properties of these cells caused by cytoskeletal disruption. Bar = 20 μm

Recent work in our laboratory has focused on the contribution of the cytoskeleton during the process of osteogenic (bone forming cell) differentiation. We have exploited the effect of cytochalasin on the actin cytoskeleton of human mesenchymal stem cells (hMSCs) to study the changes of the cytoskeleton as a stem cell changes from a somewhat "generic" cell to a cell with a specific purpose, namely to form bone tissue. Fluorescent rhodamine-phalloidin staining (specific for F-actin[180]) reveals the intense transformation of the actin cytoskeleton upon osteogenic differentiation; see Figure 2. The actin cytoskeleton becomes robust and ordered upon osteogenic differentiation. It can be speculated that the natural forces that MSCs encounter in a physiological environment do not necessitate a strong cytoskeleton. However, upon osteogenic differentiation, in which the stem cells become a part of a larger bone structure that functions to provide both form and strength, the supporting structure of the cells becomes enhanced to better enable function. The hMSCs also abandon their long, fibroblast-like, spindle shape for more of a circular shape which may better fit their role as bone forming cell. A greater effect of cytochalasin D on the actin cytoskeleton of undifferentiated hMSCs than on hMSCs differentiated down the osteogenic lineage also supports the idea that the cytoskeleton of osteogenic hMSCs is robust as compared to undifferentiated hMSCs; see Figure 2. At different concentrations of cytochalasin D, the cytoskeleton of osteogenic MSCs was less disrupted than that of undifferentiated MSCs; see Figure 3.

7.2. Fluid Flow Forces

It has been proposed that mechanical loading of bone causes fluid flow around bone cells and ultimately deformation (strain) at the cell membrane.[1] Blood in the circulation system constantly exposes endothelial cells to shear and strain forces by a fluid shear stress and endothelial cells have been proposed as sensors and regulators of vessel structure and morphogenesis.[184, 185] Erythrocytes, or red blood cells (RBCs), are constantly exposed to a shear stress by the fluid portion of the blood as they move through the circulatory system.

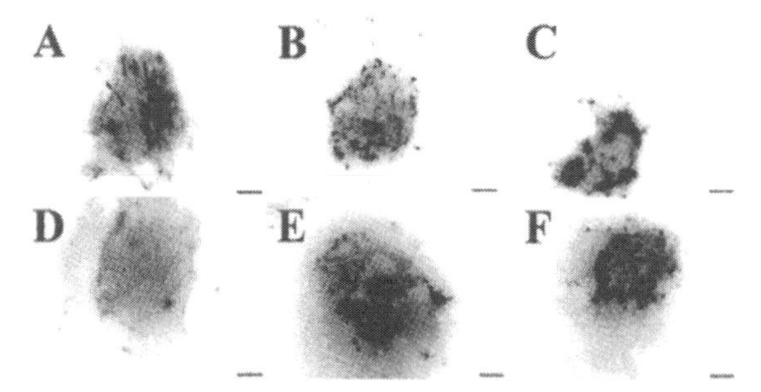

Figure 3. At varying doses of cytochalasin D (0.1 μg: A, D; 0.5 μg: B, E; 1.0 μg: C, F), the actin cytochalasin of non-differentiated MSCS (A-C) was affected more than that of osteogenic MSCs (D-F). Bar = 5 μm

7.2.1. Parallel-Plate Flow Chamber Methodology

Slides are placed in a parallel-plate flow chamber at 80% confluence (this is the optimum cell density to allow cells to change shape, to better visualize cell morphology and for more precise measurement with AFM). The apparatus used in our laboratory has been described previously.[76] In brief, the system consists of two cylindrical glass reservoirs, one above the other, with a parallel-plate flow chamber connected to them. The distance between the upper reservoir and the return outlet of the flow chamber drives the fluid flow through the chamber by the hydrostatic pressure head created. A constant pressure is created by continuous pumping of culture medium between the lower to upper reservoir. The flow chamber is a machine-milled polycarbonate plate, a rectangular gasket on top of the bottom polycarbonate plate, a polycarbonate slide with the attached cell layer on top of the gasket, and a top machine-milled polycarbonate plate. These will be held together by ten nuts and bolts spaced around the outside of the plates. The polycarbonate plate has two holes through which medium enters and exits the channel. All parts of the flow loop apparatus are washed, dried, assembled, and autoclaved prior to the experiment.

The wall shear stress on the cell monolayer can be calculated assuming Newtonian fluid and parallel-plate geometry:

$$\tau = \frac{6Q\mu}{bh^2} \tag{4}$$

where Q is the flow rate (cm^2/s); μ is the viscosity of medium used in experiment (ca. 0.01 dynes/cm$^{2)}$; h is the channel height (0.022 cm); b is the slit width (2.5 cm); and τ is the wall shear stress (dynes/cm^2).

7.2.2. *Examples of Cytoskeletal Importance During Fluid Flow*

When periosteal fibroblasts, osteocytes, and a mixture of osteoblasts and osteocytes are subjected to a pulsatile fluid flow (PFF) of 0.5 Pa at 5 Hz,[186] secretion of prostaglandins E_2 and I_2 (hormonal second messengers) both increase, the greatest increase being seen in the population of osteocytes alone. When cytochalasin B was used to disrupt the actin cytoskeleton, this response was completely blocked. This showed that in bone cells, the actin cytoskeleton is involved in the response to PFF and therefore, the cytoskeleton may play an important role in bone mechanotransduction.

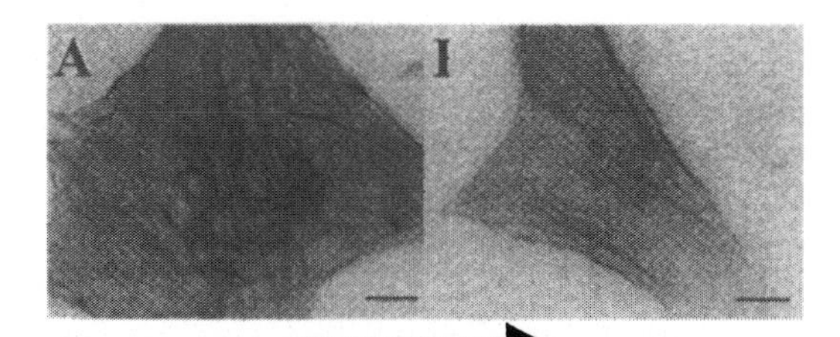

Figure 4. Human MSCs not exposed (A) or exposed (B), to a fluid flow stimulus of 9 dynes/cm^2 for 24 hr. Note the general reorganization of the actin cytoskeleton of the cell exposed to a fluid flow. Arrow indicates direction of flow. Bar = 5 μm

Osteopontin (OPN) has long been known to be an important marker of bone formation.[187, 188] It is an Arginine-Glycine-Aspartic Acid (Arg-Gly-Asp; RGD)-containing protein and was initially described as one of the important noncollagenous proteins that accumulates in the extracellular matrix (ECM) of bone in many verterbrates.[189, 190] Expression of osteopontin is a response to mechanical stimulation of bone cells *in vitro*.[191, 192] The importance of the actin cytoskeleton in relation to osteopontin expression was demonstrated by applying a biaxial strain on a custom device similar to a Flexcell at 0.25 Hz (physiologically equivalent to a fast walk) for single 2 hr periods to embryonic chick osteoblasts. The mRNA expression of the *opn* gene was unregulated in response to the mechanical strain as compared to the unstrained cells. The role of two components of the cytoskeleton, namely the actin filaments and microtubules, was studied by exposure of the cells to cytoskeleton disrupting drugs; cytochalasin D for actin filaments and colchicine for microtubules. While the expression of *opn* was prevented by the disruption of the actin cytoskeleton, there was no effect of the disruption of the microtubule cytoskeleton. Endothelial cells have been shown to respond to flow by a change in shape and alignment.[193-200] The significance of microfilaments in these processes has been demonstrated.[201] The cytoskeleton is necessary for EC adherence under shear conditions, as

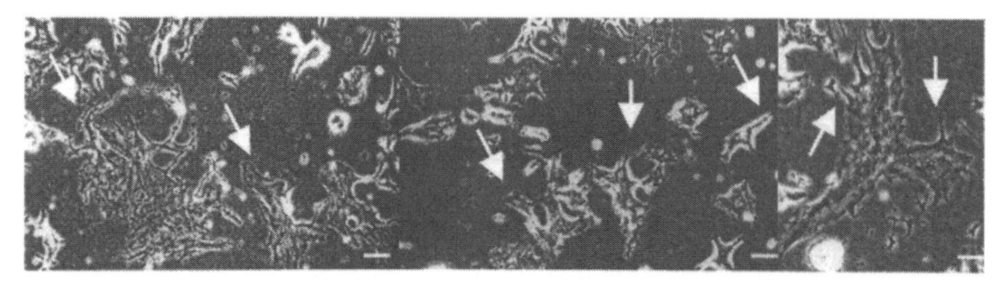

Figure 5. Actin disrupted human MSCs subjected to fluid flow induced shear stress. Note the spiky processes of the cells (denoted by arrows). Bar = 5 μm.

ECs treated with cytochalasin B detach at high shear rates.[201] ET-1 is produced by endothelial and vascular smooth muscle cells[202] and stimulates proliferation and migration of endothelial cells.[203-206] The importance of the cytoskeleton in relation to the endothelin-1 (ET-1) gene expression was demonstrated by a work by Malek et al.[207] Disruption of the actin cytoskeleton of bovine aortic endothelial cells (BAECs) was accomplished using cytochalasin D. ET-1 mRNA levels of BAECs exposed to cytochalasin D decreased in a time- and dose-dependent manner. In a follow up study, the group subjected the BAECs to a 6 h steady laminar shear stress of 20 dyn/cm^2 using a cone-plate viscometer; however, found no downregulation of ET-1 mRNA levels. They concluded from this fact that the microfilament network may be critical for resisting shear force; however, is not required for sustained ET-1 mRNA downregulation by shear.

Recently, the AFM has become a valuable tool for studying the effects of fluid flow on the membrane of a cell. A decrease in peak cell height, as determined by AFM, resulted along with EC elongation and orientation with flow.[32] Fluorescence microscopy illustrated that stress fibers of F-actin bundles mainly formed in the central portion of the sheared cells, with random formation in non-control cells. Using Finite Element Analysis, (FEM) experimental results, and the Hertz model, it was predicted that the elastic modulus of the cells would increase upon exposure to shear stress.

Erythrocytes as well are a good candidate for mechanical stimulation studies as they are exposed to supraphysiological forces by artificial organs developed to assist the circulatory system. The deformation ability of erythrocytes is important in their task of correctly delivering oxygen to the body. Even in low shear stresses, this deformation ability is changed and platelets become activated.[208] On the whole cell scale, it is difficult to delineate any morphological changes from a non-stressed to a stressed erythrocyte. However, with AFM, it becomes apparent that the fine structures on the cell surface (lipid membrane and associated membrane proteins) increase greatly as a result of shear stress by a cone and plate viscometer, as demonstrated by an increase in nano-protrusions.[209] However, surface roughness decreases with application of a fluid force. The magnitude of both of these properties increases with increasing times.

Recent work in our laboratory has focused on the effects that a fluid flow has on the cytoskeleton of human MSCs, as illustrated in Figure 4. Preliminary results have pointed to a general reorganization of the actin cytoskeleton in response to a constant fluid flow with shear stress of 9 dynes/cm^2 for 24 hours via a parallel-plate flow chamber apparatus with hydrostatic pressure driven by a flow loop. This reorganization may indicate an acceleration of differentiation of the MSCs towards an osteogenic lineage, as a reorganization of the actin cytoskeleton has been noted in osteogenic differentiation of human

MSCs; see Figure 2. Also, actin cytoskeleton disrupted cells develop spiky processes when they are subject to a fluid flow with shear stress of 9 dynes/cm^2 for a 24 hours, as illustrated in Figure 5.

8. PHYSICAL CHARACTERIZATION OF LIVING CELLS AND SURROUNDING ENVIRONMENT WITH AFM

8.1. Determination of Surface Roughness and Nanomechanical Properties of Bone Marrow-Derived Mesenchymal Stem Cells

Here we discuss recent work performed in our laboratory to assay the mechanical properties of bone marrow derived mesenchymal stem cells using atomic force microscopy. These cells were grown on poly-D-lysine-treated glass cover slips (12 mm diameter) and incubated with Dulbecco's minimum essential medium (DMEM) supplemented with 10% fetal bovine serum (FBS) and 1% antibiotic-antimycotic solution in 95% air/5% CO_2 at 37°C. To induce osteogenesis or chondrogenesis, MSCs were subcultured identically as control, with the addition of osteogenic or chondrogenic supplements (as described in section 3.1).

8.2. Imaging of MSCs with AFM

Both topographic and force spectroscopy images were obtained upon nanoindentation with cantilever probes in contact mode using a Nanoscope IIIa atomic force microscope (Veeco Digital Instruments, Inc., Santa Barbara, CA). Cantilevers with a nominal force constant of k = 0.06 N/m and oxide-sharpened Si_3N_4 tips were used to apply nanoindentation against MSC membrane surface. Scan rates were set at 1 Hz for topographic imaging and 14 Hz for force spectroscopy while scan size was set at 10 μm^2,[146, 147] so that indentation remained within the boundary of average-sized MSC. The radius of the curvature of the scanning tips was approximately 20 nm. A fluid cell without was used for imaging that contained 50 μl of fresh serum-free culture medium to retain the MSCs in a hydrated environment. AFM cantilever probes were driven by a piezoscanner to indent identified cell membrane surfaces (see Figure 1A) and yielded both topographic and force-volume images in real time.

8.3. Results and analysis

The mean surface roughness for each cell was derived from three different and randomly selected 10 μm^2 scanning fields of the topographic height images and was determined by using the following equation:

$$R = \frac{\sum_{i=1}^{N} \left| Z_i - Z_{cp} \right|}{N} \tag{5}$$

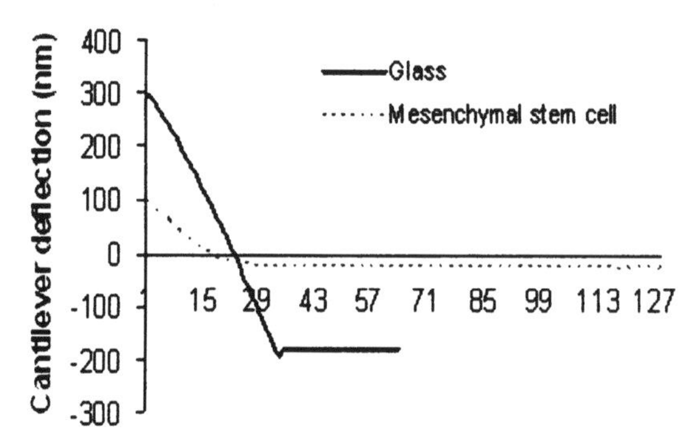

Figure 5. Force vs. displacement curves (trace only). The same z-piezo displacement results in a smaller cantilever deflection on the membrane surface of the mesenchymal stem cell in comparison to the hard glass surface because of the elastic indentation.

where Z_{cp} is the Z value of the center plane, Z_i is the current Z value, and N is the number of points within a given area (Digital Instruments Technical Note Version 3.0 No. 004-130-000).[146, 147] Similarly, the average Young's modulus for each MSC was derived from individual calculations of three randomly selected points on the membrane surface within the 10 μm^2 scanning field, using the previously discussed Hertz model.[143, 146, 147, 178]

The membrane surface of MSCs showed substantial height variation as peaks and valleys upon AFM nanoindentation in randomly selected10 μm^2 membrane areas (see Figure 1B), in sharp contrast to the smooth surface of the glass coverslip carrier, without medium (see Figure 1C) or with medium (see Figure 1D). The mean surface roughness of cell membrane calculated from topographic images of 10 randomly selected MSCs was 5.88±1.53 nm in the Z dimension. Cantilever deflection and AFM tip movement in the Z dimension have a linear relationship when the tip contacts a hard surface such as a glass cover slip (solid line in Figure 5), but are non-linear on a soft surface such as the MSC membrane (dotted line in Figure 5). Accordingly, indentation for the MSC mem-

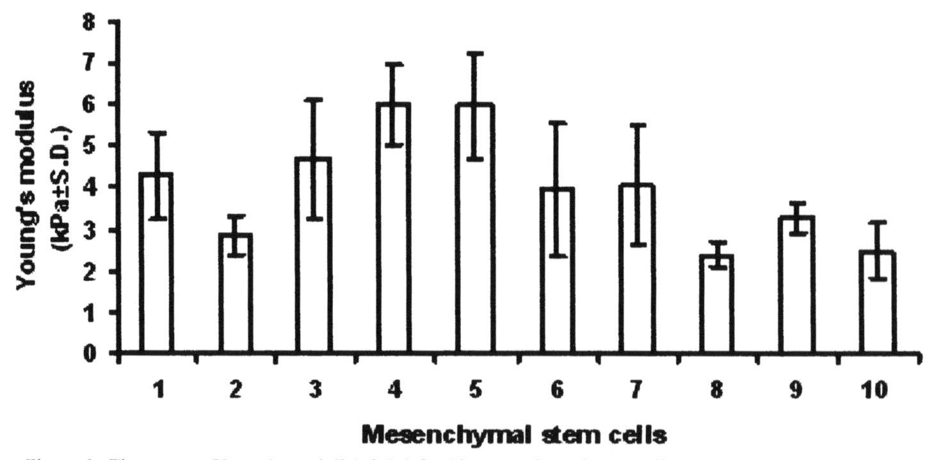

Figure 6. The average Young's moduli (±S.D.) for 10 mesenchymal stem cells.

brane surface was determined by subtracting the deflection for the membrane from the deflection for the glass. These parameters were used in the Hertz model to derive Young's modulus for each MSC, with a distribution shown in Figure 6. The mean Young's modulus calculated for control MSCs was 3.99±0.95 kPa. The mean Young's modulus calculated for osteogenic MSCs was 54.11±27.23 kPa. The mean Young's modulus for chondrogenic MSCs was 42.05±18.79 kPa.

This example shows that the elastic properties of mesenchymal stem cells, similar to other types of cells, can be measured with AFM and that living MSCs have a high tolerance to relatively large nanoindentational forces (up to 3 nN) acting on a limited region of less than 100 nm represented by the contact area between the AFM tip and the MSCs membrane surface. The Young's moduli for MSCs ranging from 2.4 to 5.98 kPa identified here are within the same range reported for other cell types such as fibroblasts,[159] and chondrocytes.[92] The lower Young's modulus of non-differentiated MSCs indicates that the membrane of MSCs is less rigid than other terminally differentiated cells such as fibroblasts and chondrocytes. This also demonstrates a change in Young's modulus accompanies a change in the cell's function upon osteogenic and chondrogenic differentiation of the cells This increase in modulus, indicating an increase in cell membrane stiffness, reiterates the idea that the structure of the cell undergoes a transformation upon differentiation. As bone and cartilage cells, the differentiated stem cells are subjected to extreme amounts and cycles of forces which may require increase cell stiffness.

AFM appears appropriate for studying both topographic and nanomechanical properties of surface structures including mesenchymal stem cells. The nanoelastic properties of MSCs in the range of 2.4 to 5.98 kPa may serve as baseline for additional characterization of their physical properties. Our recent work of encapsulating MSC-differentiated chondrogenic and osteogenic cells in hydrogel polymers for a tissue-engineered mandibular condyle further indicates the need to fully understand biophysical properties of MSCs.[9] Current work includes inducing osteogenic and chondrogenic differentiation of MSCs and comparing the nanomechanical properties of MSC-differentiated cells.

9. CONCLUDING REMARKS

We have described examples from our laboratory of using atomic force microscopy to image the nanomechanical properties of a cell membrane and discussed a cell nanostructure, the cytoskeleton, and its relationship to cell differentiation and mechanics. The vast potential of AFM in the study of biological nanostructures is clear and many exciting studies are being performed in multiple laboratories. Here we have an example of the tremendous impact of physical science in our understanding of biological structures.

10. REFERENCES

1. S. Weinbaum, S. C. Cowin and Y. Zeng, A model for the excitation of osteocytes by mechanical loading-induced bone fluid shear stresses, *J. Biomech.* **27**, 339-60 (1994).
2. L. You, S. C. Cowin, E. Schaffer and S. Weinbaum, A model for strain amplification in the actin cytoskeleton of osteocytes due to fluid drag on pericellular matrix, *J. Biomech.* **34**, 1375-86 (2001).
3. A. Alhadlaq, R. V. Patel, D. Lennon, A. I. Caplan, T. Ayalin and J. J. Mao, 3064 (2002).
4. K. L. Duncan, Transduction of mechanical strain in bone, *ASGSB Bull.* **8**, 49-62 (1995).

5. N. X. Chen, K. D. Ryder, F. M. Pavalko, C. H. Turner, D. B. Burr, J. Qiu and K. L. Duncan, Ca(2+) regulates fluid shear-induced cytoskeletal reorganization and gene expression in osteoblasts, *Am. J. Physiol. Cell Physiol.* **278**, 989-97 (2000).
6. J. J. Mao, Mechanobiology of craniofacial sutures, *J. Dent. Res.* **81**, 810-6 (2002).
7. J. M. Collins, K. Ramamoorthy, A. Da Silveira, P. A. Patston and J. J. Mao, Microstrain in intramembranous bones induces altered gene expression of MMP1 and MMP2 in the rat, *J. Biomech.* (2004).
8. R. A. Kopher and J. J. Mao, Suture growth modulated by the oscillatory component of micromechanical strain, *J. Bone Miner. Res.* **18**, 521-8 (2003).
9. A. Alhadlaq and J. J. Mao, Tissue-engineered Neogenesis of Human-shaped Mandibular Condyle from Rat Mesenchymal Stem Cells, *J. Dent. Res.* **82**, 951-6 (2003).
10. A. I. Caplan, The mesengenic process, *Clin. Plast. Surg.* **21**, 429-35 (1994).
11. M. Owen, Marrow stromal stem cells, *J. Cell Sci. Suppl.* **10**, 63-76 (1988).
12. J. N. Beresford, Osteogenic stem cells and the stromal system of bone and marrow, *Clin. Orthop.*, 270-80 (1989).
13. H. Ohgushi, V. M. Goldberg and A. I. Caplan, Heterotopic osteogenesis in porous ceramics induced by marrow cells, *J. Orthop. Res.* **7**, 568-78 (1989).
14. J. H. Bennett, C. J. Joyner, J. T. Triffitt and M. E. Owen, Adipocytic cells cultured from marrow have osteogenic potential, *J. Cell Sci.* **99 (Pt 1)**, 131-9 (1991).
15. M. W. Long, J. A. Robinson, E. A. Ashcraft and K. G. Mann, Regulation of human bone marrow-derived osteoprogenitor cells by osteogenic growth factors, *J. Clin. Invest.* **95**, 881-7 (1995).
16. S. A. Azizi, D. Stokes and B. J. Augelli, Engraftment and migration of human bone marrow stromal cells implanted in the brains of albino rats--similarities to astrocyte grafts, *P. Natl. Acad. Sci. USA* **95**, 3908-13 (1998).
17. G. Ferrari, G. Cusella-De Angelis and M. Coletta, Muscle regeneration by bone marrow-derived myogenic progenitors, *Science* **279**, 1528-30 (1998).
18. R. G. Young, D. L. Butler and W. Weber, Use of mesenchymal stem cells in a collagen matrix for Achilles tendon repair, *J. Orthop. Res.* **16**, 406-13 (1998).
19. Friedenstein AJ. Determined and inducible osteogenic precursor cells. In: Hard Tissue Growth, Repair and Remineralization, Ciba Fdn Symp, North-Holland: Elsevier-Excerpa Medica, 1973, p. 169-185.
20. A. Friedenstein and A. I. Kuralesova, Osteogenic precursor cells of bone marrow in radiation chimeras, *Transplantation* **12**, 99-108 (1971).
21. J. Goshima, V. M. Goldberg and A. I. Caplan, The origin of bone formed in composite grafts of porous calcium phosphate ceramic loaded with marrow cells, *Clin. Orthop.*, 274-83 (1991).
22. I. Martin, A. Muraglia, G. Campanile, R. Cancedda and R. Quarto, Fibroblast growth factor-2 supports ex vivo expansion and maintenance of osteogenic precursors from human bone marrow, *Endocrinology* **138**, 4456-62 (1997).
23. M. F. Pittenger, A. M. Mackay, S. C. Beck, R. K. Jaiswal, R. Douglas, J. D. Mosca, M. A. Moorman, D. W. Simonetti, S. Craig and D. R. Marshak, Multilineage potential of adult human mesenchymal stem cells, *Science* **284**, 143-7 (1999).
24. A. I. Caplan and S. P. Bruder, Mesenchymal stem cells: building blocks for molecular medicine in the 21st century, *Trends Mol. Med.* **7**, 259-64 (2001).
25. J. E. Dennis, J. P. Carbillet, A. I. Caplan and P. Charbord, The STRO-1+ marrow cell population is multipotential, *Cells Tissues Organs* **170**, 73-82 (2002).
26. N. Quirici, D. Soligo, P. Bossolasco, F. Servida, C. Lumini and G. L. Deliliers, Isolation of bone marrow mesenchymal stem cells by anti-nerve growth factor receptor antibodies, *Exp. Hematol.* **30**, 783-91 (2002).
27. E. A. Jones, S. E. Kinsey, A. English, R. A. Jones, L. Straszynski, D. M. Meredith, A. F. Markham, A. Jack, P. Emery and D. McGonagle, Isolation and characterization of bone marrow multipotential mesenchymal progenitor cells, *Arthritis Rheum.* **46**, 3349-60 (2002).
28. E. J. Caterson, L. J. Nesti, K. G. Danielson and R. S. Tuan, Human marrow-derived mesenchymal progenitor cells: isolation, culture expansion, and analysis of differentiation, *Mol. Biotechnol.* **20**, 245-56 (2002).
29. B. Alberts, A. Johnson, J. Lewis, M. Raff, K. Roberts and P. Walter, *Molecular Biology of the Cell* (Garland Science, New York, 2002).
30. A. J. Maniotis, C. S. Chen and D. E. Ingber, Demonstration of mechanical connections between integrins, cytoskeletal filaments, and nucleoplasm that stabilize nuclear structure, *Proc. Natl. Acad. Sci. U S A* **94**(3), 849-854 (1997).

31. C. Rotsch and M. Radmacher, Drug-induced changes of cytoskeletal structure and mechanics in fibroblasts: an atomic force microscopy study, *Biophys. J.* **78**, 520-35 (2000).
32. T. Ohashi, Y. Ishii, Y. Ishikawa, T. Matsumoto and K. Sato, Experimental and numerical analyses of local mechanical properties measured by atomic force microscopy for sheared endothelial cells, *Biomed. Mater. Eng.* **12**, 319-27 (2002).
33. J. Lee, A. Ishihara and K. Jacobson, How do cells move along surfaces?, *Trends Cell. Biol.* **3**, 366-70 (1993).
34. T. P. Stossel, On the crawling of animal cells, *Science* **260**, 1086-94 (1993).
35. M. R. Bubb, A. M. Senderowicz, E. A. Sausville, K. L. Duncan and E. D. Korn, Jasplakinolide, a cytotoxic natural product, induces actin polymerization and competitively inhibits the binding of phalloidin to F-actin, *J. Biol. Chem.* **269**, 14869-71 (1994).
36. J. A. Cooper, Effects of cytochalasin and phalloidin on actin, *J. Cell Biol.* **105**, 1473-8 (1987).
37. I. Spector, N. R. Shochet, D. Blasberger and Y. Kashman, Latrunculins--novel marine macrolides that disrupt microfilament organization and affect cell growth: I. Comparison with cytochalasin D, *Cell Motil. Cytoskeleton* **13**, 127-44 (1989).
38. N. O. Petersen, W. B. McConnaughey and E. L. Elson, Dependence of locally measured cellular deformability on position on the cell, temperature, and cytochalasin B, *P. Natl. Acad. Sci. USA* **79**, 5327-31 (1982).
39. G. C. Reilly, J. D. Currey and A. E. Goodship, Exercise of young thoroughbred horses increases impact strength of the third metacarpal bone, *J. Orthop. Res.* **15**, 862-8 (1997).
40. E. L. Batson, G. C. Reilly, J. D. Currey and D. S. Balderson, Postexercise and positional variation in mechanical properties of the radius in young horses, *Equine. Vet. J.* **32**, 95-100 (2000).
41. D. R. Carter and G. S. Beaupre, *Skeletal Function and Form: Mechanobiology of Skeletal Development, Aging, and Regeneration* (Cambridge University Press, Cambridge, 2001).
42. S. B. Arnaud, D. J. Sherrard, N. Maloney, R. T. Whalen and P. Fung, Effects of 1-week head-down tilt bed rest on bone formation and the calcium endocrine system, *Aviat. Space Environ. Med.* **63**, 14-20 (1992).
43. C. L. Donaldson, S. B. Hulley, J. M. Vogel, R. S. Hattner, J. H. Bayers and D. E. McMillan, Effect of prolonged bed rest on bone mineral, *Metabolism* **19**, 1071-84 (1970).
44. R. D. Roer and R. M. Dillaman, Bone growth and calcium balance during simulated weightlessness in the rat, *J. Appl. Physiol.* **68**, 13-20 (1990).
45. G. D. Whedon, Disuse osteoporosis: physiological aspects, *Calcif. Tissue Int.* **36 Suppl 1**, 146-50 (1984).
46. V. S. Schneider and J. McDonald, Skeletal calcium homeostasis and countermeasures to prevent disuse osteoporosis, *Calcif. Tissue Int.* **36 Suppl 1**, 151-44 (1984).
47. A. E. Tami, M. B. Schaffler and M. L. K. Tate, Probing the tissue to subcellular level structure underlying bone's molecular sieving function, *Biorheology* **40**(6), 577-590 (2003).
48. K. M. Reich, C. V. Gay and J. A. Frangos, Fluid shear stress as a mediator of osteoblast cyclic adenosine monophosphate production, *J. Cell. Physiol.* **143**, 100-4 (1990).
49. M. V. Hillsley and J. A. Frangos, Bone tissue engineering: the role of interstitial fluid flow, *Biotechnol. Bioeng.* **43**, 573-81 (1994).
50. G. C. Reilly, T. R. Haut, C. E. Yellowley, H. J. Donahue and C. R. Jacobs, Fluid flow induced PGE2 release by bone cells is reduced by glycocalyx degradation whereas calcium signals are not, *Biorheology* **40**, 591-603 (2003).
51. G. N. Bancroft, V. I. Sikavitsas, D. J. van den, T. L. Sheffield, C. G. Ambrose, J. A. Jansen and A. G. Mikos, Fluid flow increases mineralized matrix deposition in 3D perfusion culture of marrow stromal osteoblasts in a dose-dependent manner, *Proc. Natl. Acad. Sci. U. S. A* **99**(20), 12600-12605 (2002).
52. M. E. Gomes, V. I. Sikavitsas, E. Behravesh, R. L. Reis and A. G. Mikos, Effect of flow perfusion on the osteogenic differentiation of bone marrow stromal cells cultured on starch-based three-dimensional scaffolds, *J. Biomed. Mater. Res.* **67A**(1), 87-95 (2003).
53. V. I. Sikavitsas, G. N. Bancroft, H. L. Holtorf, J. A. Jansen and A. G. Mikos, Mineralized matrix deposition by marrow stromal osteoblasts in 3D perfusion culture increases with increasing fluid shear forces, *Proc. Natl. Acad. Sci. U. S. A* **100**(25), 14683-14688 (2003).
54. D. J. van den, G. N. Bancroft, V. I. Sikavitsas, P. H. Spauwen, J. A. Jansen and A. G. Mikos, Flow perfusion culture of marrow stromal osteoblasts in titanium fiber mesh, *J. Biomed. Mater. Res.* **64A**(2), 235-241 (2003).
55. J. A. Cooper, M. Hewison and P. M. Stewart, Glucocorticoid activity, inactivity and the osteoblast, *J. Endocrinol.* **163**, 159-64 (1999).

56. C. Maniatopoulos, J. Sodek and A. H. Melcher, Bone formation in vitro by stromal cells obtained from bone marrow of young adult rats, *Cell Tissue Res.* **254**, 317-30 (1988).
57. C. G. Bellows, J. E. Aubin and J. N. Heersche, Initiation and progression of mineralization of bone nodules formed in vitro: the role of alkaline phosphatase and organic phosphate, *Bone Miner.* **14**, 27-40 (1991).
58. C. H. Chung, E. E. Golub, E. Forbes, T. Tokuoka and I. M. Shapiro, Mechanism of action of beta-glycerophosphate on bone cell mineralization, *Calcif. Tissue Int.* **51**, 305-11 (1992).
59. C. H. Chung, D. Z. Liu, S. Y. Wang and S. S. Wang, Enhancement of the growth of human endothelial cells by surface roughness at nanometer scale, *Biomaterials* **24**, 4655-61 (2003).
60. A. M. Mackay, S. C. Beck, J. M. Murphy, F. P. Barry, C. O. Chichester and M. F. Pittenger, Chondrogenic differentiation of cultured human mesenchymal stem cells from marrow, *Tissue Eng.* **4**, 415-28 (1998).
61. J. U. Yoo, T. S. Barthel, K. Nishimura, L. Solchaga, A. I. Caplan, V. M. Goldberg and B. Johnstone, The chondrogenic potential of human bone-marrow-derived mesenchymal progenitor cells, *J. Bone Joint Surg. Am.* **80**, 1745-57 (1998).
62. R. Cancedda, F. Descalzi Cancedda and P. Castagnola, Chondrocyte differentiation, *Int. Rev. Cytol.* **159**, 265-358 (1995).
63. B. M. Spiegelman and H. Green, Cyclic AMP-mediated control of lipogenic enzyme synthesis during adipose differentiation of 3T3 cells, *Cell* **24**, 503-10 (1981).
64. A. Giganti and E. Friederich, The actin cytoskeleton as a therapeutic target: state of the art and future directions, *Prog. Cell Cycle Res.* **5**, 511-525 (2003).
65. K. N. Bhalla, Microtubule-targeted anticancer agents and apoptosis, *Oncogene* **22**(56), 9075-9086 (2003).
66. J. Tannenbaum, Approaches to the molecular biology of cytochalasin action, *Front. Biol.* **46**, 521-559 (1978).
67. S. S. Brown and J. A. Spudich, Cytochalasin inhibits the rate of elongation of actin filament fragments, *J. Cell Biol.* **83**(3), 657-662 (1979).
68. S. L. Brenner and E. D. Korn, Substoichiometric concentrations of cytochalasin D inhibit actin polymerization. Additional evidence for an F-actin treadmill, *J. Biol. Chem.* **254**(20), 9982-9985 (1979).
69. D. C. Lin, K. D. Tobin, M. Grumet and S. Lin, Cytochalasins inhibit nuclei-induced actin polymerization by blocking filament elongation, *J. Cell Biol.* **84**(2), 455-460 (1980).
70. M. D. Flanagan and S. Lin, Cytochalasins block actin filament elongation by binding to high affinity sites associated with F-actin, *J. Biol. Chem.* **255**(3), 835-838 (1980).
71. A. Mozo-Villarias and B. R. Ware, Distinctions between mechanisms of cytochalasin D activity for Mg2+- and K+-induced actin assembly, *J. Biol. Chem.* **259**(9), 5549-5554 (1984).
72. K. Maruyama and K. Tsukagoshi, Effects of KCl, MgCl2, and CaCl2 concentrations on the monomer-polymer equilibrium of actin in the presence and absence of cytochalasin D, *J. Biochem. (Tokyo)* **96**(3), 605-611 (1984).
73. J. H. Hartwig and T. P. Stossel, Cytochalasin B and the structure of actin gels, *J. Mol. Biol.* **134**(3), 539-553 (1979).
74. K. Maruyama, J. H. Hartwig and T. P. Stossel, Cytochalasin B and the structure of actin gels. II. Further evidence for the splitting of F-actin by cytochalasin B, *Biochim. Biophys. Acta* **626**(2), 494-500 (1980).
75. L. A. Selden, L. C. Gershman and J. E. Estes, A proposed mechanism of action of cytochalasin D on muscle actin, *Biochem. Biophys. Res. Commun.* **95**(4), 1854-1860 (1980).
76. J. A. Frangos, L. V. McIntire and S. G. Eskin, Shear stress induced stimulation of mammalian cell metabolism, *Biotechnol. Bioeng.* **32**, 1053-1060 (1988).
77. J. A. Frangos, S. G. Eskin, L. V. McIntire and C. L. Ives, Flow effects on prostacyclin production by cultured human endothelial cells, *Science* **227**, 1477-9 (1985).
78. C. R. Jacobs, C. E. Yellowley, B. R. Davis, Z. Zhou, J. M. Cimbala and H. J. Donahue, Differential effect of steady versus oscillating flow on bone cells, *J. Biomech.* **31**, 969-76 (1998).
79. J. You, G. C. Reilly, X. Zhen, C. E. Yellowley, Q. Chen, H. J. Donahue and C. R. Jacobs, Osteopontin gene regulation by oscillatory fluid flow via intracellular calcium mobilization and activation of mitogen-activated protein kinase in MC3T3-E1 osteoblasts, *J. Biol. Chem.* **276**, 13365-71 (2001).
80. S. M. McCormick, P. A. Whitson, H. W. Wu and L. V. McIntire, Shear stress differentially regulates PGHS-1 and PGHS-2 protein levels in human endothelial cells, *Ann. Biomed. Eng.* **28**, 824-33 (2000).
81. S. M. McCormick, S. R. Frye, S. G. Eskin, C. L. Teng, C. M. Lu, C. G. Russell, K. K. Chittur and L. V. McIntire, Microarray analysis of shear stressed endothelial cells, *Biorheology* **40**, 5-11 (2003).

82. S. M. McCormick, S. G. Eskin, L. V. McIntire, C. L. Teng, C. M. Lu, C. G. Russell and K. K. Chittur, DNA microarray reveals changes in gene expression of shear stressed human umbilical vein endothelial cells, *P. Natl. Acad. Sci. USA* **98**(16), 8955-8960 (2001).
83. L. V. McIntire and S. G. Eskin, Mechanical and biochemical aspects of leukocyte interactions with model vessel walls, *Kroc. Found. Ser.* **16**, 209-219 (1984).
84. G. S. Worthen, L. A. Smedly, M. G. Tonnesen, D. Ellis, N. F. Voelkel, J. T. Reeves and P. M. Henson, Effects of shear stress on adhesive interaction between neutrophils and cultured endothelial cells, *J. Appl. Physiol* **63**(5), 2031-2041 (1987).
85. K. Sakai, M. Mohtai and Y. Iwamoto, Fluid shear stress increases transforming growth factor beta 1 expression in human osteoblast-like cells: modulation by cation channel blockades, *Calcif. Tissue Int.* **63**(6), 515-520 (1998).
86. A. M. Malek, J. Zhang, J. Jiang, S. L. Alper and S. Izumo, Endothelin-1 gene suppression by shear stress: pharmacological evaluation of the role of tyrosine kinase, intracellular calcium, cytoskeleton, and mechanosensitive channels, *J. Mol. Cell Cardiol.* **31**(2), 387-399 (1999).
87. H. Morawietz, R. Talanow, M. Szibor, U. Rueckschloss, A. Schubert, B. Bartling, D. Darmer and J. Holtz, Regulation of the endothelin system by shear stress in human endothelial cells, *J. Physiol* **525 Pt 3**, 761-770 (2000).
88. C. A. Simmons, S. Matlis, A. J. Thornton, S. Chen, C. Y. Wang and D. J. Mooney, Cyclic strain enhances matrix mineralization by adult human mesenchymal stem cells via the extracellular signal-regulated kinase (ERK1/2) signaling pathway, *J. Biomech.* **36**, 1087-96 (2003).
89. F. A. Weyts, B. Bosmans, R. Niesing, J. P. van Leeuwen and H. Weinans, Mechanical control of human osteoblast apoptosis and proliferation in relation to differentiation, *Calcif. Tissue Int.* **72**, 505-12 (2003).
90. B. P. Jena and H. Horber, *Atomic Force Microscopy in Cell Biology* (Academic Press, San Diego, 2002).
91. S. Kasas and A. Ikai, A method for anchoring round shaped cells for atomic force microscope imaging, *Biophys. J.* **68**, 1678-80 (1995).
92. D. L. Bader, T. Ohashi, M. M. Knight, D. A. Lee and K. Sato, Deformation properties of articular chondrocytes: a critique of three separate techniques, *Biorheology* **39**, 69-78 (2002).
93. E. Henderson, Atomic force microscopy of living cells, *Prog. Surf. Sci.* **46**, 39-60 (1994).
94. V. J. Morris, A. R. Kirby and A. P. Gunning, *Atomic Force Microscopy for Biologists* (Imperial College Press, London, 1999).
95. A. Boisen, O. Hansen and S. Bouwstra, AFM probes with directly fabricated tips, *J. Micromech. Microeng.* **6**, 58-62 (1996).
96. Y. G. Kuznetsov, A. J. Malkin and A. McPherson, Atomic force microscopy studies of living cells: visualization of motility, division, aggregation, transformation, and apoptosis, *J. Struct. Biol.* **120**, 180-91 (1997).
97. M. Radmacher, Measuring the elastic properties of biological samples with the AFM, *IEEE Eng. Med. Bio. Mag.* **16**, 47-57 (1997).
98. S. S. Wong, E. Joselevich, A. T. Woolley, C. L. Cheung and C. M. Lieber, Covalently functionalized nanotubes as nanometre-sized probes in chemistry and biology, *Nature* **394**, 52-5 (1998).
99. C. L. Cheung, J. H. Hafner and C. M. Lieber, Carbon nanotube atomic force microscopy tips: direct growth by chemical vapor deposition and application to high-resolution imaging, *P. Natl. Acad. Sci. USA* **97**, 3809-13 (2000).
100. J. H. Hafner, C. L. Cheung, A. T. Woolley and C. M. Lieber, Structural and functional imaging with carbon nanotube AFM probes, *Prog. Biophys. Mol. Biol.* **77**, 73-110 (2001).
101. Sethuraman A, Stroscio MA and Dutta M. Potential application of carbon nanotubes in bioengineering. edited by Stroscio MA. 2004.
102. A. T. Woolley, C. L. Cheung, J. H. Hafner and C. M. Lieber, Structural biology with carbon nanotube AFM probes, *Chem. Biol.* **7**, 193-204 (2000).
103. Y. Yang, C. Y. Wang and D. A. Erie, Quantitative characterization of biomolecular assemblies and interactions using atomic force microscopy, *Methods* **29**, 175-187 (2003).
104. S. Iwabuchii, T. Mori, K. Ogawa, K. Sato, M. Saito, Y. Morita, T. Ushiki and E. Tamiya, Atomic force microscope-based dissection of human metaphase chromosomes and high resolutional imaging by carbon nanotube tip, *Arch. Histol. Cytol.* **65**, 473-9 (2002).
105. J. L. Alonso and W. H. Goldmann, Feeling the forces: atomic force microscopy in cell biology, *Life Sci.* **72**, 2553-60 (2003).
106. J. H. Hoh and C. A. Schoenenberger, Surface morphology and mechanical properties of MDCK monolayers by atomic force microscopy, *J. Cell Sci.* **107 (Pt 5)**, 1105-14 (1994).

107. M. Melling, D. Karimian-Teherani, M. Behnam and S. Mostler, Morphological study of the healthy human oculomotor nerve by atomic force microscopy, *Neuroimage* **20**, 795-801 (2003).
108. J. A. Dvorak, The application of atomic force microscopy to the study of living vertebrate cells in culture, *Methods* **29**, 86-96 (2003).
109. H.-X. You and L. Yu, Atomic force microscopy imaging of living cells: progress, problems and prospects, *Methods Cell Sci.* **21**, 1-17 (1999).
110. F. M. Ohnesorge, J. K. Horber, W. Haberle, C. P. Czerny, D. P. Smith and G. Binnig, AFM review study on pox viruses and living cells, *Biophys. J.* **73**(4), 2183-94 (1997).
111. Y. F. Dufrene, Application of atomic force microscopy to microbial surfaces: from reconstituted cell surface layers to living cells, *Micron* **32**, 153-65 (2001).
112. D. Fotiadis, S. Scheuring, S. A. Muller, A. Engel and D. J. Muller, Imaging and manipulation of biological structures with the AFM, *Micron* **33**, 385-97 (2002).
113. A. Engel and D. J. Muller, Observing single biomolecules at work with the atomic force microscope, *Nat. Struct. Biol.* **7**, 715-8 (2000).
114. M. Radmacher, R. W. Tillamnn, M. Fritz and H. E. Gaub, From molecules to cells: imaging soft samples with the atomic force microscope, *Science* **257**, 1900-5 (1992).
115. L. Chang, T. Kious, M. Yorgancioglu, D. Keller and J. Pfeiffer, Cytoskeleton of living, unstained cells imaged by scanning force microscopy, *Biophys. J.* **64**, 1282-6 (1993).
116. E. Henderson, P. G. Haydon and D. S. Sakaguchi, Actin filament dynamics in living glial cells imaged by atomic force microscopy, *Science* **257**, 1944-6 (1992).
117. H. Haga, M. Sasaki, K. Kawabata, E. Ito, T. Ushiki and T. Sambongi, Elasticity mapping of living fibroblasts by AFM and immunofluorescence observation of the cytoskeleton, *Ultramicroscopy* **82**, 253-8 (2000).
118. H. W. Wu, T. Kuhn and V. T. Moy, Mechanical properties of L929 cells measured by atomic force microscopy: effects of anticytoskeletal drugs and membrane crosslinking, *Scanning* **20**, 389-97 (1998).
119. F. Braet, R. De Zanger, W. Kalle, A. Raap, H. Tanke and E. Wisse, Comparative scanning, transmission and atomic force microscopy of the microtubular cytoskeleton in fenestrated liver endothelial cells, *Scanning Microsc. Suppl.* **10**, 225-35 (1996).
120. A. M. Collinsworth, S. Zhang, W. E. Kraus and G. A. Truskey, Apparent elastic modulus and hysteresis of skeletal muscle cells throughout differentiation, *Am. J. Physiol. Cell Physiol.* **283**, 1219-27 (2002).
121. U. G. Hofmann, C. Rotsch, W. J. Parak and M. Radmacher, Investigating the cytoskeleton of chicken cardiocytes with the atomic force microscope, *J. Struct. Biol.* **119**, 84-91 (1997).
122. Y. Yamane, H. Shiga, H. Haga, K. Kawabata, K. Abe and E. Ito, Quantitative analyses of topography and elasticity of living and fixed astrocytes, *J. Electron Microsc. (Tokyo)* **49**, 463-71 (2000).
123. F. Braet, R. De Zanger, C. Seynaeve, M. Baekeland and E. Wisse, A comparative atomic force microscopy study on living skin fibroblasts and liver endothelial cells, *J. Electron Microsc. (Tokyo)* **50**, 283-90 (2001).
124. F. Braet, C. Seynaeve, R. De Zanger and E. Wisse, Imaging surface and submembranous structures with the atomic force microscope: a study on living cancer cells, fibroblasts and macrophages, *J. Microsc.* **190 (Pt 3)**, 328-38 (1998).
125. F. Liu, J. Burgess, H. Mizukami and A. Ostafin, Sample preparation and imaging of erythrocyte cytoskeleton with the atomic force microscopy, *Cell Biochem. Biophys.* **38**, 251-70 (2003).
126. M. Radmacher, M. Fritz, C. M. Kacher, J. P. Cleveland and P. K. Hansma, Measuring the viscoelastic properties of human platelets with the atomic force microscope, *Biophys. J.* **70**(1), 556-567 (1996).
127. M. Walch, U. Ziegler and P. Groscurth, Effect of streptolysin O on the microelasticity of human platelets analyzed by atomic force microscopy, *Ultramicroscopy* **82**, 259-67 (2000).
128. A. Simon, T. Cohen-Bouhacina, M. C. Porte, J. P. Aime, J. Amedee, R. Bareille and C. Baquey, Characterization of dynamic cellular adhesion of osteoblasts using atomic force microscopy, *Cytometry* **54A**, 36-47 (2003).
129. H. Zimmermann, R. Hagedorn, E. Richter and G. Fuhr, Topography of cell traces studied by atomic force microscopy, *Eur. Biophys. J.* **28**, 516-25 (1999).
130. D. Ricci and M. Grattarola, Scanning force microscopy on live cultured cells: imaging and force-versus-distance investigations, *J. Microsc.* **176 (Pt 3)**, 254-61 (1994).
131. D. Ricci, M. Tedesco and M. Grattarola, Mechanical and morphological properties of living 3T6 cells probed via scanning force microscopy, *Microsc. Res. Tech.* **36**, 165-71 (1997).

132. C. Le Grimellec, E. Lesniewska, M. C. Giocondi, E. Finot and J. P. Goudonnet, Simultaneous imaging of the surface and the submembraneous cytoskeleton in living cells by tapping mode atomic force microscopy, *C. R. Acad. Sci. III* **320**, 637-43 (1997).
133. O. Chumakova, A. Liopo, S. A. Chizhik, V. V. Tayurskaya, L. L. Gerashchenko and O. Y. Komkov, Effects of ethanol and acetaldehyde on isolated nerve ending membranes: study by atomic-forced microscopy, *Bull. Exp. Biol. Med.* **130**, 921-4 (2000).
134. A. Liopo, O. Chumakova, I. Zavodnik, A. Andreyeva, M. Bryszewska and S. A. Chizhik, The response of the neuronal membrane to acetaldehyde treatment, *Cell Mol. Biol. Lett.* **6**, 265-9 (2001).
135. T. Wakatsuki, B. Schwab, N. C. Thompson and E. L. Elson, Effects of cytochalasin D and latrunculin B on mechanical properties of cells, *J. Cell Sci.* **114**, 1025-36 (2001).
136. M. C. Giocondi, V. Vie, E. Lesniewska, J. P. Goudonnet and C. Le Grimellec, In situ imaging of detergent-resistant membranes by atomic force microscopy, *J. Struct. Biol.* **131**, 38-43 (2000).
137. P. E. Milhiet, M. C. Giocondi and C. Le Grimellec, AFM Imaging of Lipid Domains in Model Membranes, *Sci. World J.* **3**, 59-74 (2003).
138. S. Yamashina and O. Katsumata, Structural analysis of red blood cell membrane with an atomic force microscope, *J. Electron Microsc. (Tokyo)* **49**, 445-51 (2000).
139. D. J. Muller, G. M. Hand, A. Engel and G. E. Sosinsky, Conformational changes in surface structures of isolated connexin 26 gap junctions, *Embo J.* **21**, 3598-607 (2002).
140. M. Gad, A. Itoh and A. Ikai, Mapping cell wall polysaccharides of living microbial cells using atomic force microscopy, *Cell Biol. Int.* **21**, 697-706 (1997).
141. L. Scheffer, A. Bitler, E. Ben-Jacob and R. Korenstein, Atomic force pulling: probing the local elasticity of the cell membrane, *Eur. Biophys. J.* **30**, 83-90 (2001).
142. M. McElfresh, E. Baesu, R. Balhorn, J. Belak, M. J. Allen and R. E. Rudd, Combining constitutive materials modeling with atomic force microscopy to understand the mechanical properties of living cells, *P. Natl. Acad. Sci. USA* **99 Suppl 2**, 6493-7 (2002).
143. W. F. Heinz and J. H. Hoh, Spatially resolved force spectroscopy of biological surfaces using the atomic force microscope, *Trends Biotechnol.* **17**(4), 143-150 (1999).
144. H. G. Hansma and J. H. Hoh, Biomolecular imaging with the atomic force microscope, *Annu. Rev. Biophys. Biomol. Struct.* **23**, 115-39 (1994).
145. A. L. Weisenhorn, P. Maivald, H. Butt and H. G. Hansma, Measuring adhesion, attraction, and repulsion between surfaces in liquids with an atomic-force microscope, *Phys. Rev. B Con. Matt.* **45**, 11226-11232 (1992).
146. K. Hu, P. Radhakrishnan, R. V. Patel and J. J. Mao, Regional structural and viscoelastic properties of fibrocartilage upon dynamic nanoindentation of the articular condyle, *J. Struct. Biol.* **136**(1), 46-52 (2001).
147. R. V. Patel and J. J. Mao, Microstructural and elastic properties of the extracellular matrices of the superficial zone of neonatal articular cartilage by atomic force microscopy, *Front. Biosci.* **8**, 18-25 (2003).
148. J. M. Gere, *Mechanics of Materials* (Brooks/Cole, Pacific Grove, 2001).
149. G. T. Charras and M. A. Horton, Determination of cellular strains by combined atomic force microscopy and finite element modeling, *Biophys. J.* **83**, 858-79 (2002).
150. K. Sato, K. Nagayama, N. Kataoka, M. Sasaki and K. Hane, Local mechanical properties measured by atomic force microscopy for cultured bovine endothelial cells exposed to shear stress, *J. Biomech.* **33**, 127-35 (2000).
151. M. Lekka, P. Laidler, D. Gil, J. Lekki, Z. Stachura and A. Z. Hrynkiewicz, Elasticity of normal and cancerous human bladder cells studied by scanning force microscopy, *Eur. Biophys. J.* **28**, 312-6 (1999).
152. S. G. Shroff, D. R. Saner and R. Lal, Dynamic micromechanical properties of cultured rat atrial myocytes measured by atomic force microscopy, *Am. J. Physiol.* **269**, 286-92 (1995).
153. A. H. Swihart, J. M. Mikrut, J. B. Ketterson and R. C. Macdonald, Atomic force microscopy of the erythrocyte membrane skeleton, *J. Microsc.* **204**, 212-25 (2001).
154. C. Rotsch, F. Braet, E. Wisse and M. Radmacher, AFM imaging and elasticity measurements on living rat liver macrophages, *Cell Biol. Int.* **21**, 685-96 (1997).
155. F. Braet, C. Rotsch, E. Wisse and M. Radmacher, Comparison of fixed and living endothelial cells by atomic force microscopy, *Appl. Phys. A* **66**, S575-S578 (1997).
156. H. Miyazaki and K. Hayashi, Atomic force microscopic measurement of the mechanical properties of intact endothelial cells in fresh arteries, *Med. Biol. Eng. Comput.* **37**, 530-6 (1999).
157. A. B. Mathur, G. A. Truskey and W. M. Reichert, Atomic force and total internal reflection fluorescence microscopy for the study of force transmission in endothelial cells, *Biophys. J.* **78**, 1725-35 (2000).

158. C. Rotsch, K. Jacobson and M. Radmacher, Dimensional and mechanical dynamics of active and stable edges in motile fibroblasts investigated by using atomic force microscopy, *P. Natl. Acad. Sci. USA* **96**, 921-6 (1999).
159. K. Sinniah, J. Paauw and J. Ubels, Investigating live and fixed epithelial and fibroblast cells by atomic force microscopy, *Curr. Eye Res.* **25**, 61-8 (2002).
160. J. Domke, S. Dannohl, W. J. Parak, D. J. Muller, W. K. Aicher and M. Radmacher, Substrate dependent differences in morphology and elasticity of living osteoblasts investigated by atomic force microscopy, *Colloids Surf. B. Biointerfaces* **19**, 367-379 (2000).
161. G. T. Charras, P. P. Lehenkari and M. A. Horton, Atomic force microscopy can be used to mechanically stimulate osteoblasts and evaluate cellular strain distributions, *Ultramicroscopy* **86**, 85-95 (2001).
162. M. Sugawara, Y. Ishida and H. Wada, Local mechanical properties of guinea pig outer hair cells measured by atomic force microscopy, *Hear. Res.* **174**, 222-9 (2002).
163. C. Le Grimellec, M. C. Giocondi, M. Lenoir, M. Vater, G. Sposito and R. Pujol, High-resolution three-dimensional imaging of the lateral plasma membrane of cochlear outer hair cells by atomic force microscopy, *J. Comp. Neurol.* **451**, 62-9 (2002).
164. W. H. Goldmann, R. Galneder, M. Ludwig, W. Xu, E. D. Adamson, N. Wang and R. M. Ezzell, Differences in elasticity of vinculin-deficient F9 cells measured by magnetometry and atomic force microscopy, *Exp. Cell Res.* **239**, 235-42 (1998).
165. E. L. Elson, Cellular mechanics as an indicator of cytoskeletal structure and function, *Annu. Rev Biophys. Biophys. Chem.* **17**, 397-430 (1988).
166. J. A. Theriot and T. J. Mitchison, Actin microfilament dynamics in locomoting cells, *Nature* **352**, 126-31 (1991).
167. T. D. Pollard and G. G. Borisy, Cellular motility driven by assembly and disassembly of actin filaments, *Cell* **112**, 453-65 (2003).
168. S. W. Grill, J. Howard, E. Schaffer, E. H. Stelzer and A. A. Hyman, The distribution of active force generators controls mitotic spindle position, *Science* **301**, 518-21 (2003).
169. J. M. Scholey, I. Brust-Mascher and A. Mogilner, Cell division, *Nature* **422**, 746-52 (2003).
170. U. Dammer, O. Popescu, P. Wagner, D. Anselmetti, H. J. Guntherodt and G. N. Misevic, Binding strength between cell adhesion proteoglycans measured by atomic force microscopy, *Science* **267**, 1173-5 (1995).
171. M. Benoit, Cell adhesion measured by force spectroscopy on living cells, *Methods Cell Biol.* **68**, 91-114 (2002).
172. C. J. Weijer, Visualizing signals moving in cells, *Science* **300**, 96-100 (2003).
173. M. Schliwa and G. Woehlke, Molecular motors, *Nature* **422**, 759-65 (2003).
174. R. C. May and L. M. Machesky, Phagocytosis and the actin cytoskeleton, *J. Cell Sci.* **114**, 1061-77 (2001).
175. R. E. Harrison and S. Grinstein, Phagocytosis and the microtubule cytoskeleton, *Biochem. Cell Biol.* **80**, 509-15 (2002).
176. T. Y. Lee and A. I. Gotlieb, Microfilaments and microtubules maintain endothelial integrity, *Microsc. Res. Tech.* **60**, 115-27 (2003).
177. Y. Yoon, K. Pitts and M. McNiven, Studying cytoskeletal dynamics in living cells using green fluorescent protein, *Mol. Biotechnol.* **21**, 241-50 (2002).
178. A. H. E, W. F. Heinz, M. D. Antonik, N. P. D'Costa, S. Nageswaran, C. A. Schoenenberger and J. H. Hoh, Relative microelastic mapping of living cells by atomic force microscopy, *Biophys. J.* **74**, 1564-78 (1998).
179. M. L. Wang, L. J. Nesti, R. Tuli, J. Lazatin, K. G. Danielson, P. F. Sharkey and R. S. Tuan, Titanium particles suppress expression of osteoblastic phenotype in human mesenchymal stem cells, *J. Orthop. Res.* **20**(6), 1175-1184 (2002).
180. E. Wulf, A. Deboben, F. A. Bautz, H. Faulstich and T. Wieland, Fluorescent phallotoxin, a tool for the visualization of cellular actin, *Proc. Natl. Acad. Sci. U S A* **76**(9), 4498-4502 (1979).
181. N. C. Zanetti and M. Solursh, Induction of chondrogenesis in limb mesenchymal cultures by disruption of the actin cytoskeleton, *J. Cell Biol.* **99**(1 Pt 1), 115-123 (1984).
182. Y. B. Lim, S. S. Kang, T. K. Park, Y. S. Lee, J. S. Chun and J. K. Sonn, Disruption of actin cytoskeleton induces chondrogenesis of mesenchymal cells by activating protein kinase C-alpha signaling, *Biochem. Biophys. Res. Commun.* **273**(2), 609-613 (2000).
183. Y. B. Lim, S. S. Kang, W. G. An, Y. S. Lee, J. S. Chun and J. K. Sonn, Chondrogenesis induced by actin cytoskeleton disruption is regulated via protein kinase C-dependent p38 mitogen-activated protein kinase signaling, *J. Cell Biochem.* **88**(4), 713-718 (2003).

184. P. F. Davies, Mechanisms involved in endothelial responses to hemodynamic forces, *Atherosclerosis* **131 Suppl**, S15-S17 (1997).
185. M. A. Gimbrone, Jr., N. Resnick, T. Nagel, L. M. Khachigian, T. Collins and J. N. Topper, Hemodynamics, endothelial gene expression, and atherogenesis, *Ann. N. Y. Acad. Sci.* **811**, 1-10 (1997).
186. N. E. Ajubi, J. Klein-Nulend, P. J. Nijweide, T. Vrijheid-Lammers, M. J. Alblas and E. H. Burger, Pulsating fluid flow increases prostaglandin production by cultured chicken osteocytes--a cytoskeleton-dependent process, *Biochem. Biophys. Res. Commun.* **225**(1), 62-68 (1996).
187. W. T. Butler, The nature and significance of osteopontin, *Connect. Tissue Res.* **23**(2-3), 123-136 (1989).
188. M. P. Mark, W. T. Butler, C. W. Prince, R. D. Finkelman and J. V. Ruch, Developmental expression of 44-kDa bone phosphoprotein (osteopontin) and bone gamma-carboxyglutamic acid (Gla)-containing protein (osteocalcin) in calcifying tissues of rat, *Differentiation* **37**(2), 123-136 (1988).
189. D. T. Denhardt and X. Guo, Osteopontin: a protein with diverse functions, *FASEB J.* **7**(15), 1475-1482 (1993).
190. L. C. Gerstenfeld, T. Uporova, S. Ashkar, E. Salih, Y. Gotoh, M. D. McKee, A. Nanci and M. J. Glimcher, Regulation of avian osteopontin pre- and posttranscriptional expression in skeletal tissues, *Ann. N. Y. Acad. Sci.* **760**, 67-82 (1995).
191. L. V. Harter, K. A. Hruska and R. L. Duncan, Human osteoblast-like cells respond to mechanical strain with increased bone matrix protein production independent of hormonal regulation, *Endocrinology* **136**(2), 528-535 (1995).
192. T. Kubota, M. Yamauchi, J. Onozaki, S. Sato, Y. Suzuki and J. Sodek, Influence of an intermittent compressive force on matrix protein expression by ROS 17/2.8 cells, with selective stimulation of osteopontin, *Arch. Oral Biol.* **38**(1), 23-30 (1993).
193. C. F. Dewey, Jr., S. R. Bussolari, M. A. Gimbrone, Jr. and P. F. Davies, The dynamic response of vascular endothelial cells to fluid shear stress, *J. Biomech. Eng* **103**(3), 177-185 (1981).
194. S. G. Eskin, C. L. Ives, L. V. McIntire and L. T. Navarro, Response of cultured endothelial cells to steady flow, *Microvasc. Res.* **28**(1), 87-94 (1984).
195. R. P. Franke, M. Grafe, H. Schnittler, D. Seiffge, C. Mittermayer and D. Drenckhahn, Induction of human vascular endothelial stress fibres by fluid shear stress, *Nature* **307**(5952), 648-649 (1984).
196. C. F. Dewey, Jr., Effects of fluid flow on living vascular cells, *J. Biomech. Eng* **106**(1), 31-35 (1984).
197. A. Remuzzi, C. F. Dewey, Jr., P. F. Davies and M. A. Gimbrone, Jr., Orientation of endothelial cells in shear fields in vitro, *Biorheology* **21**(4), 617-630 (1984).
198. M. J. Levesque and R. M. Nerem, The elongation and orientation of cultured endothelial cells in response to shear stress, *J. Biomech. Eng* **107**(4), 341-347 (1985).
199. A. R. Wechezak, R. F. Viggers and L. R. Sauvage, Fibronectin and F-actin redistribution in cultured endothelial cells exposed to shear stress, *Lab. Invest.* **53**(6), 639-647 (1985).
200. R. P. Franke, M. Grafe, U. Dauer, H. Schnittler and C. Mittermayer, Stress fibres (SF) in human endothelial cells (HEC) under shear stress, *Klin. Wochenschr.* **64**(19), 989-992 (1986).
201. A. R. Wechezak, T. N. Wight, R. F. Viggers and L. R. Sauvage, Endothelial adherence under shear stress is dependent upon microfilament reorganization, *J. Cell Physiol.* **139**(1), 136-146 (1989).
202. E. R. Levin, Endothelins, *N. Engl. J. Med.* **333**(6), 356-363 (1995).
203. M. S. Goligorsky, A. S. Budzikowski, H. Tsukahara and E. Noiri, Co-operation between endothelin and nitric oxide in promoting endothelial cell migration and angiogenesis, *Clin. Exp. Pharmacol. Physiol* **26**(3), 269-271 (1999).
204. L. Morbidelli, C. Orlando, C. A. Maggi, F. Ledda and M. Ziche, Proliferation and migration of endothelial cells is promoted by endothelins via activation of ETB receptors, *Am. J. Physiol* **269**(2 Pt 2), H686-H695 (1995).
205. E. Noiri, Y. Hu, W. F. Bahou, C. R. Keese, I. Giaever and M. S. Goligorsky, Permissive role of nitric oxide in endothelin-induced migration of endothelial cells, *J. Biol. Chem.* **272**(3), 1747-1752 (1997).
206. A. D. Wren, C. R. Hiley and T. P. Fan, Endothelin-3 mediated proliferation in wounded human umbilical vein endothelial cells, *Biochem. Biophys. Res. Commun.* **196**(1), 369-375 (1993).
207. A. M. Malek, I. W. Lee, S. L. Alper and S. Izumo, Regulation of endothelin-1 gene expression by cell shape and the microfilament network in vascular endothelium, *Am. J. Physiol.* **273**(5 Pt 1), C1764-C1774 (1997).
208. F. A. Kuypers, Red cell membrane damage, *J. Heart Valve Dis.* **7**(4), 387-395 (1998).
209. Y. Ohta, H. Okamoto, M. Kanno and T. Okuda, Atomic force microscopic observation of mechanically traumatized erythrocytes, *Artif. Organs* **26**(1), 10-17 (2002).

HAIRPIN FORMATION IN POLYNUCLEOTIDES: A SIMPLE FOLDING PROBLEM?

Anjum Ansari and Serguei V. Kuznetsov*

1. INTRODUCTION

The biological processes in cell involving DNA, e.g. replication, recombination, DNA repair and transcription, are accompanied by localized unwinding of the DNA molecule. This unwinding may occur as a result of fluctuational opening of individual base-pairs (Wartell and Benight, 1985), or can be driven by DNA binding proteins, leading to the formation of single-stranded (ss) region(s) in DNA (Lohman and Bjornson, 1996; Bianco et al., 2001). Palindromic regions of unwinded DNA can form irregular structures by base-pairing between neighboring self-complementary sequences. These structures are known as cruciforms and hairpins. The formation of hairpins and their involvement in biological processes such as replication and transcription is now well documented in both prokaryotic and eukaryotic systems (Crews et al., 1979; Wilson and von Hippel, 1995; Glucksmann-Kuis et al., 1996; Dai et al., 1997). It has been shown that DNA hairpins can serve as intermediates in genetic recombination (Lilley, 1981; Romer et al., 1984; Roth et al., 1992), and as protein recognition sites, and can regulate transcription *in vivo* (Dai et al., 1997; Dai et al., 1998). The biological and biomedical significance of hairpin structures is also associated with the fact that the formation of DNA hairpins by GC-rich triplet repeat, and inordinate expansion of these triplets during DNA replication, leads to several genetic diseases including a progressive neuromuscular disorder (Caskey et al., 1992; Gacy et al., 1995).

Short ssDNA and RNA hairpin structures are also useful drug targets because their overall shape and geometry differ significantly from regular double-stranded DNA, and

*Anjum Ansari, Department of Physics, University of Illinois at Chicago, Chicago, IL 60607. Serguei V. Kuznetsov, Department of Physics, University of Illinois at Chicago, Chicago, IL 60607.

are used in hybridization studies to investigate the effect of secondary structure on probe hybridization (Vesnaver and Breslauer, 1991; Gregorian and Crothers, 1995; Armitage et al., 1998; Bonnet et al., 1999). Short RNA hairpins are of particular interest, now, because of rapid development of a new siRNA (small interfering RNA) technology (Hamilton and Baulcombe, 1999; Dykxhoorn et al., 2003), based on recent discovery of the silencing of specific genes by double-stranded RNA (Fire et al., 1998). The fact that siRNA hairpins can silence gene expression, *in vivo*, offers the potential for gene-function determination as well as the promise for the development of therapeutic gene silencing. These studies with hairpin structures, therefore, have important implications for widespread applications in medicine (Nielsen, 1999; Cheng et al., 2003).

Unlike DNA, RNA is typically produced as a single-stranded molecule in which the hairpins represent the dominant elements of secondary structure (Varani, 1995). RNA hairpins usually have extremely high thermostability (Cheong et al., 1990; Heus and Pardi, 1991), protect mRNAs against degradation (Klausner et al., 1993; Alifano et al., 1994; Smolke et al., 2000), and serve as nucleation sites for the initiation of RNA folding (Uhlenbeck, 1990; Sclavi et al., 1998). Tertiary interactions of hairpin loops and bulges, in turn, define the three-dimensional structure of RNA molecules, which regulate its diverse activities including catalysis, ligand binding and RNA-protein recognition (Marino et al., 1995; Doudna and Doherty, 1997; Draper, 1999; Rupert et al., 2002; Lilley, 2003). The discovery that RNA molecules can function as enzymes (Cech et al., 1981) has sparked renewed interest in the problem of RNA folding, and the inevitable comparisons with the problem of protein folding (Thirumalai and Woodson, 1996).

A detailed understanding of the mechanism and control of biological process requires knowledge of the kinetics of individual steps. One of the major efforts in basic biopolymer science is directed toward an understanding of the energetics and mechanisms by which DNA or RNA molecules form their secondary and higher order structures, which involves characterizing the stability and dynamics of such structures, and determining the factors that affect their stability. In the case of RNA, in which the hairpins make up a significant part of its secondary structure, an understanding of the structural and dynamical aspects of how hairpins fold and behave is a starting point for understanding the folding of larger RNA molecules. The folding process is complex and involves parallel pathways and kinetic traps, which make the underlying energy landscape governing RNA folding rugged (Zarrinkar and Williamson, 1994; Zarrinkar et al., 1996; Pan et al., 1997; Pan and Sosnick, 1997; Sclavi et al., 1998; Zhuang et al., 2000; Thirumalai et al., 2001). This ruggedness may be more pronounced in the early stages of the RNA folding process because misfolded secondary structures in RNA are independently stable in the absence of tertiary interactions, unlike secondary structural elements in proteins. Thus, misfolded secondary structures in RNA may serve as kinetic traps that would slow down the tertiary reorganization of RNA molecules (Thirumalai, 1998; Wu and Tinoco, 1998; Isambert and Siggia, 2000; Thirumalai et al., 2001).

Kinetics measurements on secondary structure formation in ss-polynucleotides provide insights into the conformational flexibility of these chains, and the nature and strength of the intrachain interactions. In this connection, short oligonucleotides that form hairpin structures in solution represent an ideal model system to investigate the interactions that stabilize secondary structure in polynucleotides, and to test our understanding of the stability and dynamics of these structures. Owing to their small size and well-defined folds, hairpin structures are useful test systems for energy-based structure prediction algorithms, as have been developed by several groups (Turner and Sugimoto, 1988; Zuker, 1989; Jaeger et al., 1990; Jaeger et al., 1993; Mathews et al., 1999). These algorithms are widely used in predicting stable secondary structures for RNA molecules. The ability to predict the kinetics of these hairpins provides an additional constraint on the parameters that describe their free energies (Flamm et al., 2000; Zhang and Chen, 2002; Cocco et al., 2003a). Hairpins also serve as testing grounds for the growing effort in the molecular dynamics simulations of nucleic acids (Zichi, 1995; Miller and Kollman, 1997a; 1997b; Young et al., 1997; Srinivasan et al., 1998; Williams and Hall, 1999; Sarzynska et al., 2000; Zacharias, 2001; Sorin et al., 2002; Sorin et al., 2003).

The focus of this chapter is the thermodynamics and kinetics of hairpin formation in ss-polynucleotides. Although much of the review will focus on hairpins from ssDNA rather than RNA, the basic principles underlying secondary structure formation that have emerged from experiments on ssDNA are not expected to be grossly different for RNA strands. The principal difference between DNA and RNA secondary structure is that DNA forms double helix more readily (Alberts et al., 1989).

The chapter is organized as follows. Section 2 summarizes the basic kinetics equations that describe the folding/unfolding of hairpins; Section 3 reviews the experimental and theoretical/computational results on the thermodynamics and kinetics of hairpin formation, and presents the unresolved issues; and Section 4 concludes with a brief summary.

2. KINETIC DESCRIPTION OF HAIRPIN FOLDING TRANSITION

2.1. Two-State Description

At the simplest level of description, the folding of hairpins can be described in terms of a two-state chemical reaction:

$$\text{Unfolded} \underset{k_o}{\overset{k_c}{\rightleftharpoons}} \text{Hairpin} \tag{1}$$

where k_c is the rate coefficient for the closing (folding) step and k_o is the rate coefficient for the opening (unfolding) step. The ratio of the equilibrium populations at temperature T is given by

$$\frac{f_H^{eq}}{f_U^{eq}} = \frac{k_c}{k_o} = K_{eq} = \exp\left(\frac{-\Delta G}{RT}\right) \tag{2}$$

where ΔG is the free energy difference between the hairpin and the unfolded states ($\Delta G = G_H - G_U$), R is the universal gas constant, and f_H^{eq} and f_U^{eq} are the fractional populations of the hairpin and the unfolded states, respectively, at equilibrium. In a typical thermodynamic measurement, the folded/unfolded populations are measured as a function of temperature, by monitoring some spectroscopic signature, e.g. an increase in the UV absorbance at about 268 nm with increasing temperature, which reflects a decrease in the double-helical stacking of the bases (Wartell and Benight, 1985), or an increase in the fluorescence of a fluorophore attached at one end, which, in the folded state, is quenched by another label attached at the other end (Goddard et al., 2000).

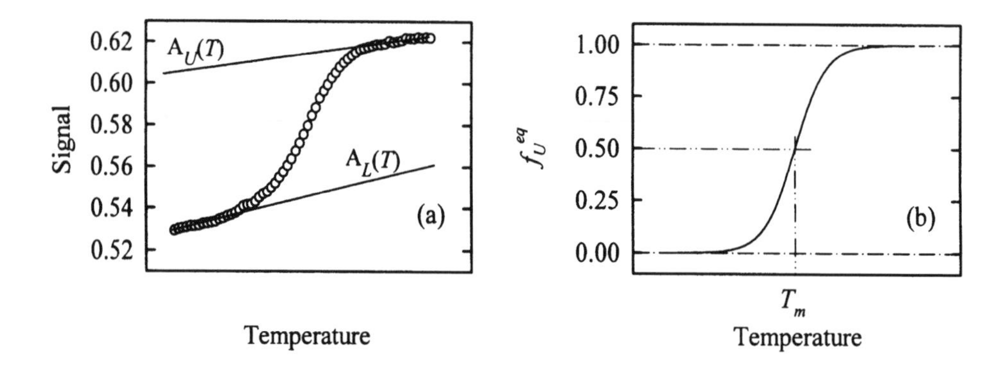

Figure 1. Melting profile of a ssDNA or RNA hairpin. (a) Absorbance versus temperature for a typical small hairpin. $A_L(T)$ and $A_U(T)$ are the lower and upper baselines, respectively. (b) The fractional population of the unfolded state f_U^{eq} versus temperature; T_m is the melting temperature.

The fractional populations are calculated from $f_U^{eq} = (A(T) - A_L(T))/(A_U(T) - A_L(T))$, where $A_L(T)$ and $A_U(T)$ are the lower and upper baselines, respectively, that reflect temperature-dependent changes in the spectroscopic signatures of the hairpin and the unfolded states, respectively. For instance, unstacking of the bases in the single-stranded regions of the hairpin and the unfolded states with increasing temperature give rise to roughly linear baselines whose absorbance increases monotonically with temperature; see Figure 1. In a two-state description, the fractional populations are described in terms of a van't Hoff expression:

$$f_U^{eq} = \frac{1}{1+K_{eq}} = \frac{1}{1+\exp\left[-\frac{\Delta H}{R}\left(\frac{1}{T}-\frac{1}{T_m}\right)\right]} \tag{3}$$

where we have expressed the free energy difference $\Delta G = \Delta H - T\Delta S = \Delta H(1-T/T_m)$. Here, ΔH is the enthalpic difference and ΔS is the entropic difference between the hairpin and the unfolded state, T_m is the melting temperature at which $\Delta G = 0$, $f_H^{eq} = f_U^{eq} = 1/2$, and the entropy change $\Delta S = \Delta H/T_m$. It is important to note here that, in Eq. (3), ΔH and ΔS are assumed to be independent of temperature, i.e. the unfolding transition is assumed to occur without any significant changes in the heat capacity ΔC_p of the system. This assumption is not quite valid for the melting of DNA, with the contribution of ΔC_p to the free energy difference ΔG found to be comparable or even larger than the contribution of ΔS to ΔG (Chalikian et al., 1999; Rouzina and Bloomfield, 1999; Williams et al., 2001). Therefore, for accurate analysis of the melting transition, the heat capacity changes should be explicitly included.

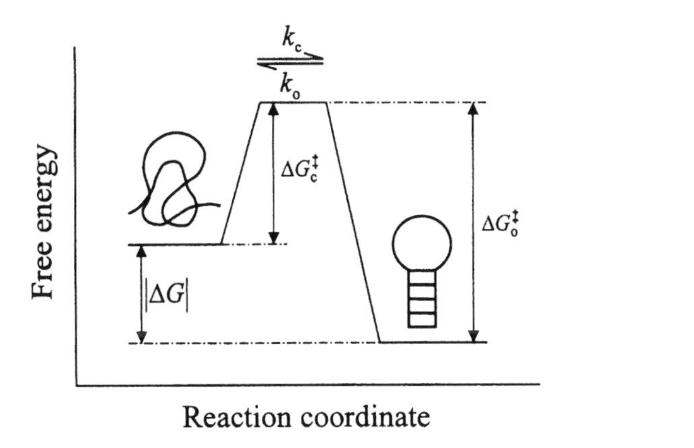

Figure 2. A schematic representation of a free energy profile for a two-state description of hairpin formation.

If the system is, initially, not at equilibrium, the fractional populations of hairpin and unfolded states will decay to their equilibrium populations with a single-exponential decay having a characteristic relaxation rate $k_r = k_o + k_c$. Thus, kinetics measurements that can be described in terms of a two-state scheme yield the sum of the opening and closing rate coefficients, whereas the equilibrium melting profiles yield the equilibrium constants and hence the ratio of the two rate coefficients as a function of temperature. In a simple description of chemical reactions, given by transition state theory, the individual rate coefficients can be written in terms of the free energy difference between the transition state and the initial state as (Steinfeld et al., 1999; Dill and Bromberg, 2003):

$$k_c = \nu \exp\left(-\frac{\Delta G_c^{\ddagger}}{RT}\right) \qquad k_o = \nu \exp\left(-\frac{\Delta G_o^{\ddagger}}{RT}\right) \tag{4}$$

where ν is the preexponential that describes an attempt frequency to go over the barrier, and $\Delta G_c^{\ddagger}$ and $\Delta G_o^{\ddagger}$ are the free energy barriers for the closing and opening steps, respectively. These barriers are illustrated in Figure 2.

2.2. Arrhenius Plots

The temperature dependence of the rate coefficients are typically described in terms of Arrhenius equations of the form

$$\begin{aligned} k_o &= \nu \exp\left(-\frac{\Delta H_o^{\ddagger}}{RT}\right)\exp\left(\frac{\Delta S_o^{\ddagger}}{R}\right) = A_o \exp\left(-\frac{\Delta H_o^{\ddagger}}{RT}\right) \\ k_c &= \nu \exp\left(-\frac{\Delta H_c^{\ddagger}}{RT}\right)\exp\left(\frac{\Delta S_c^{\ddagger}}{R}\right) = A_c \exp\left(-\frac{\Delta H_c^{\ddagger}}{RT}\right) \end{aligned} \tag{5}$$

where we have substituted the enthalpic and entropic differences into Eq. (4) for the free energy. If we assume that $\Delta H^{\ddagger}$, $\Delta S^{\ddagger}$, and the preexponential factor ν, are independent of temperature, then the slope on an Arrhenius plot of the logarithm of rates versus inverse temperature yields $-\Delta H^{\ddagger}/R$. Determining the enthalpy of the transition state relative to the beginning and end states is important for unraveling the mechanism and pathways for the transition. For example, in an elementary chemical reaction, the enthalpy of the transition state is always larger than either state. Thus, in this case, the activation enthalpy $\Delta H^{\ddagger}$ is always positive. An Arrhenius plot with a zero slope indicates that the enthalpy of the transition state is the same as that of the initial state and that the free energy barrier is entirely entropic. Finally, a negative value for the activation enthalpy is indicative of a multi-step reaction, which, when analyzed as a two-state system, yields an effective transition state which is in fact an intermediate state with *lower* enthalpy than that of the initial state. Deviations from Arrhenius behavior are to be expected if the entropy and enthalpy changes, or the preexponential in Eq. (5), are not temperature-independent. Non-Arrhenius temperature dependence observed in the folding step for proteins, most likely has contribution from both temperature dependence of the preexponentials (Bryngelson et al., 1995; Socci et al., 1996) and from a large heat capacity change upon protein folding (Schindler and Schmid, 1996; Tan et al., 1996).

2.3. Three-State Description (Nucleation and Zipping)

The next level of description of hairpin folding is the nucleation and zipping model, in which the rate-determining step is the formation of a critical nucleus, consisting of an ensemble of looped conformation with one or more base-pairs formed, to which addition of another base-pair leads to rapid zipping of the stem. At this level, hairpin folding can be described in terms of a three-state system, consisting of the open (unfolded) state, the critical nucleus, and the hairpin state, as follows:

$$\text{Unfolded} \underset{k_{nuc}^{-}}{\overset{k_{nuc}^{+}}{\rightleftharpoons}} \text{Nucleus} \underset{k_{unzip}}{\overset{k_{zip}}{\rightleftharpoons}} \text{Hairpin} \tag{6}$$

Applying steady-state conditions for the kinetic scheme in Eq. (6), under conditions for which the population of the intermediate nucleus is small compared to the unfolded and hairpin states, the overall opening and closing rates are given by:

$$k_c = \frac{k_{nuc}^{+} k_{zip}}{k_{nuc}^{-} + k_{zip}}, \qquad k_o = \frac{k_{nuc}^{-} k_{unzip}}{k_{nuc}^{-} + k_{zip}} \tag{7}$$

The size of the critical nucleus is, then, the smallest number of base-pairs for which the formation of the next adjacent base-pair is faster than the disruption of the previously formed ones, i.e., $k_{zip} >> k_{nuc}^{-}$, and the rate at which the nucleus is formed becomes the rate-determining step for the folding of the hairpin, $k_c \approx k_{nuc}^{+}$.

2.4. Zipper Model (with Misfolded States)

A more complete description of the hairpin folding is the "zipper" model that includes all microstates with partial number of base-pairs formed (Poland and Scheraga, 1970; Cantor and Schimmel, 1980). This model assumes that each nucleotide in the stem exists in one of two possible states, base-paired or open, and that all of the base-pairs occur contiguously in a single region. Initiation of nucleation can occur at any point along the stem where one or more base- pairs form to stabilize the looped conformations, and the stem grows, or "zips", from there. In an extension of the simplest scheme, the single-stranded chain can also get transiently trapped in "non-native" loops with mismatched stems, as in Figure 3, which do not lead to the complete zipping of the stem. Therefore, in the zipper model with misfolded states, hairpin formation would require several attempts in which the non-native interactions have to be broken until the correct nucleation, which leads to the complete zipping of the stem, occurs. Transient trapping in misfolded states can lead to non-Arrhenius behavior for the closing step (Socci et al., 1996; Ansari et al., 2001; Zhang and Chen, 2002).

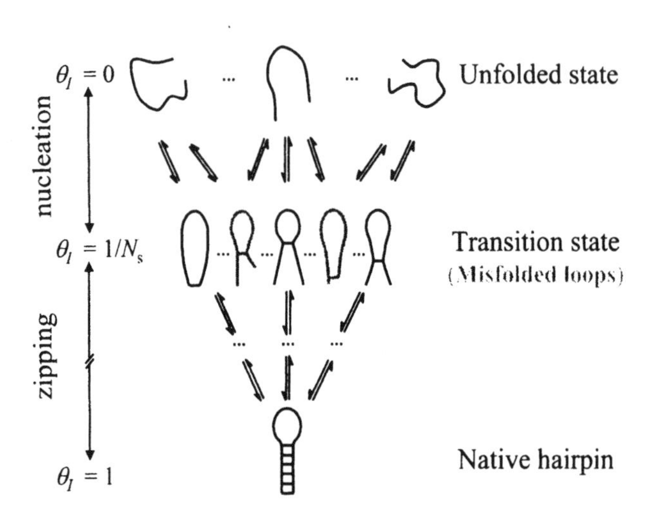

Figure 3. Zipper model (with misfolded conformations). A schematic representation of the ensemble of microstates in the unfolded, transition state, and the native state. The misfolded conformations are represented by hairpin loops with mismatched stems, which act as dead-ends in the folding process.

3. REVIEW OF EXPERIMENTAL RESULTS AND PUZZLES

The earliest measurements on the kinetics of duplex and hairpin formation were done using temperature-jump (T-jump) techniques (Cohen and Crothers, 1971; Coutts, 1971; Craig et al., 1971; Porschke and Eigen, 1971; Gralla and Crothers, 1973; Porschke, 1974b; 1974a; 1977; Chu and Tinoco, 1983; Xodo et al., 1988). The temperature of the sample was raised on microsecond time-scales using electrical pulses generated by discharging a capacitor (Eigen and de Maeyer, 1963). A modified T-jump apparatus, using a coaxial cable capacitor, has been used for some sub-microsecond measurements (Hoffman, 1971; Porschke, 1974a). An excellent discussion of some of the results from the early T-jump measurements can be found in Cantor and Schimmel (1980).

The experimental data on the kinetics of duplex formation from complementary oligonucleotides showed that the activation enthalpies for the helix formation step, for sequences containing only A·U base-pairs, are about –4 to –9 kcal/mol (Craig et al., 1971; Porschke and Eigen, 1971), and for sequences containing G·C base-pairs, the activation enthalpies are about +6 kcal/mol to +9 kcal/mol (Porschke et al., 1973). As described in the previous Section, the negative activation enthalpies observed for sequences with A·U base-pairs indicate that the helix nucleation is not a simple elementary step, but instead consists of the formation of a critical nucleus that has 2 or 3 base-pairs (Porschke, 1977). The nucleation of the helix is then the rate-determining step, followed by the rapid zipping of the stem. The positive activation enthalpies for sequences with G·C base-pairs were explained by assuming that, for these sequences, only 1 or 2 base-pairs may be sufficient to form the nucleating helix (Porschke et al., 1973).

Early estimates of the zipping rate, i.e., the rate at which a base-pair is added to an existing helix, vary widely, from very fast ($\sim 10^9\ s^{-1}$) (Spatz and Baldwin, 1965; Wetmur and Davidson, 1968) to relatively slower, between $10^7\ s^{-1}$ and $10^6\ s^{-1}$ (Craig et al., 1971; Porschke and Eigen, 1971). Porschke (1974a) made direct measurements of the zipping/unzipping rate by carrying out sub-microsecond T-jump measurements on dimers of poly(A) and poly(U) oligomers of chain lengths 14 and 18 at temperatures below the melting transition. The kinetics measurements revealed two distinct processes, one occurring with a time constant of about 0.2 μs, independent of the oligomer concentration, and a much slower relaxation occurring with a time constant of a few seconds. The slow component is the overall helix-to-coil transition, while the fast component was assigned to the unzipping at the ends. A kinetic zipper model was used to estimate the rate coefficient for the zipping step to be $\sim 8\times10^6\ s^{-1} \approx (125\ ns)^{-1}$ at 25°C (Porschke, 1974a).

Hairpin formation in ssDNA or RNA chains requires the formation of a loop stabilized by a few base-pairs as the nucleation step, followed by zipping as in duplex formation. Because of the close proximity of the two ends of the ss-chain, hairpins are expected to form on time-scales considerably faster than duplex formation. The early T-jump measurements on short self-complementary oligomers revealed that hairpins with 4-6 bases in the loop and less than 10 bases in the stem form on time-scales of tens of microseconds (Coutts, 1971; Gralla and Crothers, 1973; Porschke, 1974b). To form hairpins, the ss-chain has to overcome an entropic barrier in forming a loop, which is countered by the stabilizing free energy of a few base-pairs. To understand the time-scales for forming hairpins requires an estimation of the time-scales for forming the critical nucleus, which in turn requires reliable estimates of the free energy cost of loop formation.

It was recognized quite early that the free energy of loop formation in ss-polynucleotides deviates from the simple estimates of entropic costs expected for a random coil model, especially for loop sizes smaller than about 10 nucleotides, presumably from favorable stacking interactions of bases within small loops (Vallone et al., 1999). However, there is considerable uncertainty in the estimates of the enthalpic contribution to loop closure, obtained from the thermodynamic analysis of melting profiles of hairpins, ranging from ~21 kcal/mol (Uhlenbeck et al., 1973) to ~11 kcal/mol (Porschke, 1974b) for an $A_6C_6U_6$ hairpin, and ~0 kcal/mol for an $A_4GC_5U_4$ hairpin (Gralla and Crothers, 1973). The very large enthalpic cost for loop closure for the $A_6C_6U_6$ hairpin was explained as arising from the unstacking of cytosine residues in the poly(C) strand in order to form the loop (Uhlenbeck et al., 1973; Porschke, 1974b). However, that conclusion seems inconsistent with essentially no enthalpic cost reported by Gralla and Crothers (1973).

Kinetics measurements on the formation of hairpins impose much more stringent constraints on the possible estimates of thermodynamic parameters, and are therefore indispensable for accurate estimates of the free energies that stabilize secondary structure in ssDNA and RNA. The early T-jump measurements, however, also showed quite large variation in the activation enthalpies obtained from the temperature dependence of the measured rates, ranging from –22 kcal/mol for the closing step of a hairpin fragment $GGGCU_3GCCC$ from $tRNA^{Ser}$ (Coutts, 1971) to ~2.5 kcal/mol for the hairpin $A_6C_6U_6$ (Porschke, 1974b), suggesting a sequence dependence to the free energy of loop formation as well as to the size of the critical nucleus. Based on a comparison of the estimated enthalpy for forming the first base-pair from thermodynamic measurements on $A_6C_6U_6$, and the measured activation enthalpy for the hairpin formation step from kinetics measurements on that hairpin, Porschke argued that a stable nucleus that leads to zipping is formed only after the formation of the fourth A·U base-pair (Porschke, 1974b).

In recent years there has been a surge in the investigation of hairpin kinetics using a variety of new experimental tools, such as fluctuation correlation spectroscopy (FCS) (Bonnet et al., 1998; Goddard et al., 2000; Wallace et al., 2000); laser T-jump measurements (Ansari et al., 2001; Shen et al., 2001), and single-molecule techniques (Deniz et al., 1999; Grunwell et al., 2001; Liphardt et al., 2001).

Libchaber and co-workers have carried out a series of elegant measurements on the kinetics of conformational fluctuations in ssDNA hairpin-loops by FCS (Bonnet et al., 1998; Goddard et al., 2000). They attached a fluorophore and a quencher at either end of their oligonucleotide sequence, and the state of the molecule, whether hairpin (closed) or unfolded (open), was monitored by the intensity of the fluorescence. In the open state the molecule is fluorescent because the fluorophore and the quencher are far apart, whereas in the closed state the fluorescence is quenched. They monitored the time-scales for fluctuations between the open and the closed states by analyzing the autocorrelation function of the fluctuations in the fluorescent signal. The sequences of DNA hairpins investigated by Libchaber and co-workers were 5'-CCCAA$(X)_N$TTGGG-3', where X was either T or A, and the size of the loop (N) varied from N=12 to N=30 for the poly(dT) loop and from N=8 to N=30 for the poly(dA) loop.

The primary results from Libchaber's group are: (i) closing times depend on the sequence and length of the loop, whereas the opening times are insensitive to the loop composition; (ii) the closing times scale with the length of the loops as $\sim L^{2.6}$ for poly(dT) loops and as $\sim L^{3.2}$ for poly(dA) loops; (iii) closing times for poly(dA) loops are about 10 times slower than for poly(dT) loops at 20°C; and (iv) the activation enthalpies for the closing step increase nearly linearly for poly(dA) loops, from ~5 kcal/mol for loops with 8 bases to >15 kcal/mol for loops with 30 bases, whereas for the poly(dT) loops the activation enthalpies decrease slightly with the loop size (Bonnet et al., 1998; Goddard et al., 2000).

Kinetics measurements on another DNA hairpin 5'-GGGTT$(A)_{30}$AACCC-3', whose stem sequence is complementary to that of one of Libchaber's sequence, have been performed by Klenerman and co-workers, also using FCS techniques (Wallace et al., 2000; 2001). One difference between the two sets of FCS measurements is that in Libchaber's set-up the fluorescence of the excited label is quenched upon contact with the second label, whereas in Klenerman's set-up, the fluorescence labels attached at the two ends of the hairpin stem are donor-acceptor pair for fluorescence resonance energy transfer (FRET), and the intensity of the donor changes as the two ends come closer, but without necessarily making direct contact. Another difference is the method by which they subtract the contribution from the diffusion of the DNA molecules in and out of their observation volume to their intensity fluctuation measurements (Wallace et al., 2000). However, the results of their measurements are quite strikingly different and as yet unresolved. Libchaber's group reports single-exponential kinetics for temperatures ranging from ~10-50°C, as observed previously in T-jump measurements, whereas Klenerman's group observes highly nonexponential relaxation kinetics at ~20°C that they describe in terms of stretched exponentials (Wallace et al., 2000; 2001). The Klenerman group also reports non-Arrhenius temperature dependence for the opening and closing rates, and a viscosity dependence for the rates that scales nearly inversely with the solvent viscosity (Wallace et al., 2001).

Bustamante and co-workers (Liphardt et al., 2001) have used mechanical force to induce the unfolding and refolding of single RNA molecules, including a simple RNA hairpin, a molecule containing a three-helix junction, and a domain of a ribozyme. For their hairpin, which has ~22 base-pairs in the stem, approximately half of which are G·C base-pairs, and 4 bases in the loop, they find that the hairpin unfolds at a force of ~15 pN, similar to forces required to unzip DNA helices (Essevaz-Roulet et al., 1997; Rief et al., 1999; Bockelmann et al., 2002; Thomen et al., 2002). By imposing a constant force on the molecule, they were able to monitor the end-to-end distance between the two ends of the hairpin and to watch the distance hop back and forth between two values characteristic of the fully unfolded and the fully folded hairpin, with no evidence of any intermediate states. They determined the folding and unfolding rates from the average lifetimes in the two states, and found that, at the critical force for which the opening and closing rates are the same (~14 pN in the presence of 10 mM Mg^{2+}), the folding times are ~1s. The very slow folding times observed in these measurements, compared to the folding times of tens of microseconds observed in FCS and T-jump measurements for hairpins with similar loop sizes, but smaller stems 5-7 base-pairs long, has been explained as arising from the very large free energy barrier for folding in the presence of the applied force, which has been estimated to be ~10 k_BT (or ~6 kcal/mol) for their hairpin, and because of its long stem (Liphardt et al., 2001; Cocco et al., 2003a).

Schultz and co-workers (Deniz et al., 1999; Grunwell et al., 2001) have developed a single-molecule FRET measurement technique to monitor the conformational fluctuations of ssDNA hairpins immobilized on a glass surface. For their hairpin with 40 poly(dA) bases in the loop, they report closing times that are ~140ms, i.e. more than about 30-100 times longer than is predicted from scaling the measured closing times from the Libchaber group by $\sim L^3$. The long closing times in the single-molecule FRET measurements may be a result of the interactions of the hairpin with the derivatized glass surface, or perhaps another manifestation of the anomalous dependence of the dynamics of poly(dA) loops with increasing loop-size.

FCS and single-molecule measurements are limited in their time-resolution to microseconds and milliseconds, respectively. The rapid development of nanosecond laser T-jump techniques has opened up the field to investigate the dynamics of biomolecules with ~10 ns time-resolution (Williams et al., 1989; Hofrichter, 2001), while overcoming the limitation of the earlier T-jump setups that required the use of solutions of high conductivity and thus high ionic strength. Laser T-jump has been used extensively by several groups to investigate rapid events in the protein folding process such as the kinetics of formation of elementary secondary structures, α-helices and β-sheets (Munoz et al., 1997; Dyer et al., 1998; Eaton et al., 1998; Gruebele et al., 1998; Jager et al., 2001). In our laboratory, we have used laser T-jump to investigate hairpin dynamics in ssDNA (Ansari et al., 2001; Kuznetsov et al., 2001; Shen et al., 2001), as well as to investigate the dynamics of wrapping and unwrapping of ssDNA on a single-stranded binding protein (Kuznetsov et al., 2004). Our T-jump measurements on hairpin formation are consistent with single-exponential relaxation dynamics, although the current time-resolution is not sufficient to determine whether there is any missing amplitude on the sub-microsecond time-scale, see Figure 4. The rapid change in absorbance in the laser T-jump measurements has contributions from any unresolved relaxations and from an apparent change in the optical density of the sample from thermal lensing effects, which occurs on the time-scale of the T-jump (Hofrichter, 2001).

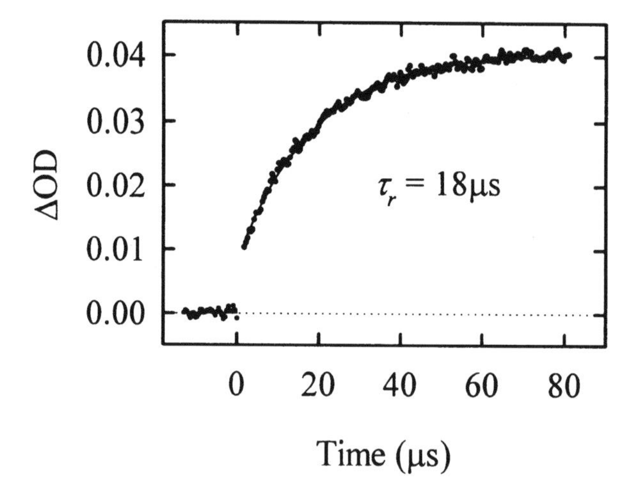

Figure 4. The change in absorbance as a function of time, for the hairpin 5'-CGGATAA(T_8)TTATCCG-3', after a T-jump from 42°C to 51°C. The kinetics are described as a single-exponential with a relaxation time of 18 μs.

The primary results from our T-jump measurements are: (i) the free energy of the hairpin relative to the unfolded state scales with the loop size with an apparent exponent of ~7, much larger than the exponent of ~1.8 expected from the entropic cost of loop formation for a semiflexible polymer; (ii) the equilibrium zipper model, which was used to calculate free energy profiles along an effective reaction coordinate, suggests that the transition state ensemble consists of looped conformations stabilized by one base-pair closing the loop; (iii) the equilibrium model predicts negative activation enthalpies of ~−9 kcal/mol for the closing step, and which are confirmed in kinetics measurements; (iv) at temperatures near T_m, the closing times for both poly(dT) loops and poly(dA) loops scale with loop size as $\sim L^2$, consistent with the scaling expected for a semiflexible polymer; (v) the opening and closing times exhibit an apparent viscosity *independence*, a conclusion that is contradictory to an earlier study on viscosity dependence by the Klenerman group (Wallace et al., 2001)

3.1. Why is hairpin formation so slow?

The nucleation step in hairpin formation requires the ss-polynucleotide to form a loop with one or more base-pairs to stabilize the loop. Models describing the characteristic time for two ends of a polymer chain to come into contact have been proposed in several theoretical studies (Wilemski and Fixman, 1974; Doi, 1975; Szabo et al., 1980; Friedman and O'Shaughnessy, 1989; Guo and Thirumalai, 1995; Podtelezhnikov and Vologodskii, 1997). An order-of-magnitude estimate for the end-to-end contact time is estimated as $\tau_D \approx <r^2>/2D_T$, where D_T is the translational diffusion coefficient of the chain and $<r^2>$ is the mean-square end-to-end distance (Winnik, 1986). We can estimate the translational diffusion coefficient from $D_T \approx k_B T/(6\pi\eta R_G)$ where η is the solvent viscosity and R_G is the radius of gyration, $R_G^2 \approx <r^2>/6$ (DeGennes, 1979). Therefore, the end-to-end contact time becomes

$$\tau_D \approx \frac{\eta <r^2>^{3/2}}{0.26 k_B T} \tag{8}$$

The result in Eq. (8) is nearly identical to the more rigorous calculation of Szabo et al. (1980) who model the dynamics of the end-to-end contact of a flexible (Gaussian) chain as diffusion in a harmonic potential well.

For a semiflexible polymer, $<r^2>$ can be written as (Landau and Lifshitz, 1980; Rivetti et al., 1998)

$$<r^2> = 2PL[1-(P/L)(1-\exp(-L/P))] \tag{9}$$

where P is the persistence length of the chain, and L is the contour length. Note that this formula predicts a stiff-rod behavior for $L<P$ and random-coil behavior for $L>>P$. For a ss-polynucleotide chain ~10 nucleotides long, and assuming an internucleotide distance of ~0.6 nm, yields L ~6nm, and $\langle r^2 \rangle \approx 12$ nm^2, where we have used a value of $P \approx 1.3$ nm (Rivetti et al., 1998). Therefore, at $T = 25°C$ and $\eta = 1$ cP, the diffusion-limited contact time is estimated to be $\tau_D \approx 40$ ns at 25°C and $D_T \approx 1.5\times10^{-6}$ cm^2/s. Wetmur and Davidson (1968) report a value of $D_T = 1.9\times10^{-8}$ cm^2/s for a ss-polynucleotide with 3.8×10^4 nucleotides. If we scale the experimentally measured value for a long ss-chain down to shorter chains using $D_T \sim 1/\sqrt{N}$, we get $D_T \approx 1.1\times10^{-6}$ cm^2/s for a strand of ~10 nucleotides, in close agreement with our crude estimate.

If the contact time between the two ends of the polymer is indeed ~40 ns, then formation of the loop cannot be the rate-determining step in hairpin formation, which occurs ~250 times slower, with hairpin closing times of ~10 μs at 25°C for hairpins with about 10 poly(dT) bases in the loop. It is well known that cyclization times for λ DNA molecule with cohesive ends are also much longer than the end-to-end contact times for a semiflexible polymer (Wang and Davidson, 1966a; 1966b; 1968). One explanation for this discrepancy was first proposed by Wang and Davidson (1966b), who argued that the rate-determining step in the joining of the two ends is the very slow chemical step of base-pair formation and not the diffusion-limited time for contact formation. They based their arguments on two observations: first, that the temperature dependence of the measured cyclization times exhibited a very large (~24 kcal/mol) activation energy; second, that the viscosity dependence of the cyclization times did not follow a simple scaling with solvent viscosity as expected for a diffusion-controlled reaction. Hairpin closing times, on the other hand, exhibit negative activation energies, especially near T_m, and therefore the chemical step of base-pair formation cannot be the rate-determining step. The viscosity dependence of the opening and closing times of a DNA hairpin is still an open question. This point is discussed further in Section 3.4.

It is of interest to compare hairpin formation in ss-polynucleotides with β-hairpin formation in polypeptides, which are also found to occur on time-scales of several microseconds (Munoz et al., 1997). Early measurements of the time-scales for loop formation in polypeptide chain under strongly denaturing conditions yielded ~40 μs for loops of ~50 residues (Hagen et al., 1996). Using a scaling of $\sim L^2$ for a semiflexible polymer of length L yields loop formation times of ~2 μs for ~10 residues long loops, which is close to the experimentally measured time of ~6 μs for the formation of a β-hairpin (Munoz et al., 1997), thus suggesting that the initiation of the loop could set the time-scale for hairpin formation.

Subsequent measurements of first contact time between two ends of Gly-rich polypeptide sequences designed to have little or no secondary structure have yielded values of ~30-100 ns for ~10 residues long loops (Bieri et al., 1999; Lapidus et al., 2000; Hudgins et al., 2002). The origin of the discrepancy between these and the earlier measurements is not clear. One suggestion is that the persistence length of the chain, which is known to be highly sequence dependent (Miller et al., 1967), was ~5 times bigger in the polypeptide chain of the denatured protein that was used in the first set of measurements compared to the designed Gly-rich sequences of the subsequent measurements (Lapidus et al., 2000). Another explanation is that the slow contact times in the early set of measurements is a result of high concentration of GdnHCl in the solution, which binds to the protein in the denatured state, and could slow down the effective diffusion coefficient of the chain (Hagen et al., 2001).

Several theoretical and computational studies of protein folding have postulated another source for the decrease in the effective diffusion coefficient of the polypeptide chain, of the form $D \sim D_0 \exp[-(\varepsilon/RT)^2]$, as a result of interactions within the chain, especially under folding conditions, which give rise to a "roughness" in the energy surface of the polypeptide (Zwanzig, 1988; Bryngelson and Wolynes, 1989; Bryngelson et al., 1995; Socci et al., 1996). Here D_0 is the intrinsic diffusion coefficient of the polymer chain and ε is the amplitude of the roughness; see Figure 5. Eaton and co-workers have postulated that even for their Gly-rich sequence especially designed to have no secondary structure, there seems to be a ~16-fold decrease in the effective diffusion coefficient of the probes attached to the two ends of the polypeptide chain, and they attributed this decrease to transient intrachain interactions (Lapidus et al., 2000).

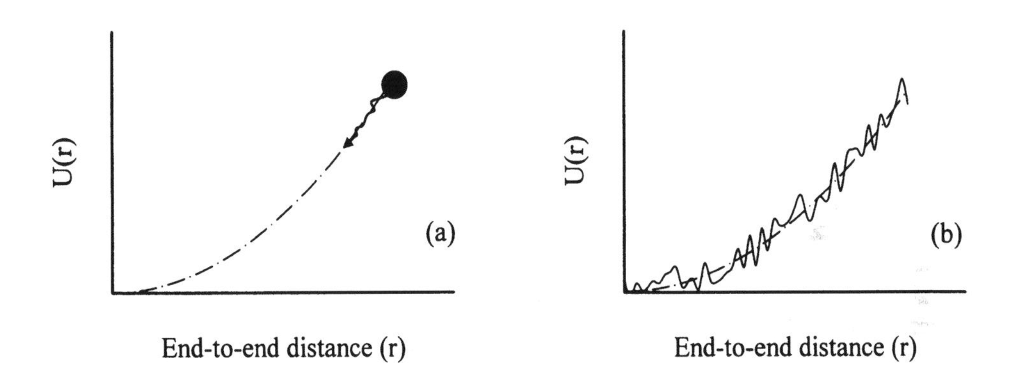

Figure 5. Diffusion in a harmonic potential. (a) For an ideal gaussian chain, the diffusion coefficient is characteristic of the relative diffusion of the two ends of the chain. (b) For chains with intrachain interactions, the harmonic potential has a roughness of amplitude ε, and the effective diffusion coefficient is reduced by a factor $\exp[-(\varepsilon/RT)^2]$.

In a series of recent papers, we proposed that such transient intrachain interactions in the unfolded state of ss-polynucleotides could lead to the slow formation of the critical nucleus for forming hairpins (Ansari et al., 2001; Kuznetsov et al., 2001; Shen et al., 2001; Ansari et al., 2002). This slowing down could arise from (i) non-native base-pairs that don't lead to a complete hairpin and that act as dead-ends during the folding process, or (ii) non-native hydrogen bonds or non-native stacking interactions (mis-stacked bases) as was suggested to explain the anomalous loop-size dependence of the hairpin closing times for hairpins with long poly(dA) loops (Ansari et al., 2002). Such a mechanism would increase the nucleation time by decreasing the effective intrachain diffusion coefficient. A characteristic roughness of only ~1.4 kcal/mol (~2.3 k_BT) would decrease the effective diffusion coefficient and increase the characteristic first contact time by a factor of ~250 at 25°C.

Two computational studies of ss-polynucleotide conformational dynamics support some of the ideas postulated above. The first study, by Zhang and Chen (2002), presents a detailed folding kinetic analysis of a 21-nucleotide RNA hairpin (9 base-pairs in the stem and 3 bases in the loop), using a statistical mechanical model that enumerates all conformations of the RNA chain with two or more contiguous (stacked) base-pairs, including all misfolded conformations. They calculate the free energy of each conformation using a statistical mechanical model for RNA thermodynamics (Chen and Dill, 2000), and using the base-pairing and stacking interactions from the RNA thermodynamics literature (Serra and Turner, 1995). The various conformations are coupled via elementary transition steps in which only one base-pair is formed or broken in any single kinetic step. The rates of transitions between the conformations are parameterized as $k_{\pm} = k_{\pm}^{0} \exp(-\Delta G_{\pm} / RT)$, where $\Delta G_{+} = T\Delta S$ is the barrier for the formation of a base-pair and is assumed to be entirely entropic, and $\Delta G_{-} = \Delta H$ is the barrier for the disruption of a base-pair, and is assumed to be the enthalpic cost of breaking the hydrogen bonding and stacking interactions. They use this model to calculate in detail the folding pathways, the relaxation kinetics, and the temperature dependence of the relaxation rates. In their model, the parameters that best describe the experimentally measured folding rates of small RNA hairpins yield ~(780 ns)$^{-1}$ for the elementary step of forming G·C base-pairs (with $k_{+} = k_{-} = 6.6 \times 10^{13}$ s^{-1}), and ~(12 ns)$^{-1}$ for forming A·U base-pairs (with $k_{+} = k_{-} = 6.6 \times 10^{12}$ s^{-1}), with ~2μs obtained for the closing time of a hairpin with 9 base-pairs in the stem and 3 bases in the loop (Shi-Jie Chen, private communication). Important results from their study include (i) a rugged energy landscape for RNA folding; (ii) folding pathways that lead to dead-ends or traps, especially at temperatures below what they define as the glass transition temperature $T_g < T_m$; and (iii) a distinctly non-Arrhenius temperature dependence for the closing rates. Near T_m, these traps are not deep; nevertheless, they could lead to slowing down of the chain dynamics.

A second study on ss-polynucleotide dynamics comes from large-scale, parallel, molecular dynamics simulation of Pande and co-workers, which involves sampling a large number of constant temperature trajectories that total more than 500 μs of simulations for an all-atom model of an RNA hairpin 5'-GGGC[GCAA]GCCU-3', with continuum representation of solvent effects (Sorin et al., 2002; Sorin et al., 2003). From their simulated trajectories, they calculate the apparent transition rates for folding by using the approximation, $k_{U \to F}^{app} = N_F / t_{total} \pm N_F^{1/2} / t_{total}$ (where N_F is the number of

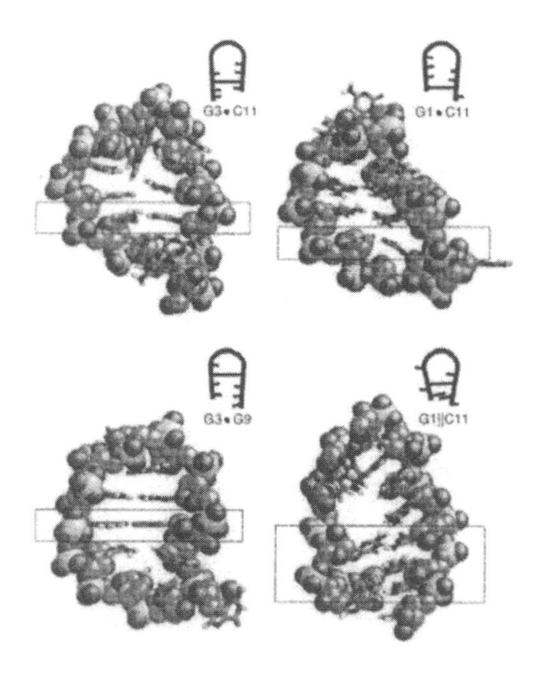

Figure 6. Members of the misfolded trap ensemble for the hairpin 5'-GGGCGCAAGCCU-3'. The figure is adapted from Sorin et al. (2003) and shows the atomistic (and schematic) pictures of the non-native interactions found in the collapsed state in their simulations.

transitions that occur from the unfolded state *U* to the folded state *F* in a total simulation time t_{total}), valid for all processes that exhibit single-exponential kinetics (Shirts and Pande, 2001; Zagrovic et al., 2001). To investigate the folding process at 300K, the simulations were started from the fully extended, denatured state. They observe at least two dominant mechanisms by which the hairpin folds, the first is a loop formation followed by zipping, and the other is a nonspecific collapse mechanism, similar to the hydrophobic collapse in proteins (Dill, 1990; Thirumalai et al., 2001). They find that a total of 21 trajectories undergo a nativelike collapse within ~175 μs, giving a collapse rate of $\sim(8\ \mu s)^{-1}$ at ~300K, which is very close to the experimentally observed hairpin closing rates. The individual conformations observed in their collapsed state show an ensemble of misfolded traps with base-pairing interactions between G3·C11 or G1·C11, hydrogen bonding interactions between G3·G9, and base-stacking interactions G1 ‖ C11, as in Figure 6. Thus, the simulations of Pande and co-workers support the notion that transient trapping can result not only from non-native base-pairing interactions, as explicitly included in the model of Zhang and Chen (2002), but also from non-native hydrogen bonding and intrastrand stacking interactions. This initial collapse and reorganization of the intrastrand contacts could then be the rate-determining step in hairpin formation.

If the time-scale for configurational diffusion to sample conformations in the unfolded state is comparable to the experimentally observed closing time for hairpins, the kinetics are expected to show deviations from single-exponential behavior. In fact, nonexponential kinetics, described in terms of stretched exponential of the form $\exp[-(kt)^{\beta}]$, have been observed by Klenerman and co-workers in the conformational fluctuations of DNA hairpins measured under equilibrium conditions at temperatures below T_m (Wallace et al., 2000). One explanation for why the Libchaber group does not see any features of nonexponential behavior for a very similar hairpin may be because in their measurements the fluorescence of their label is quenched upon contact with a label at the other end, and hence, they monitor only the open or closed state of the hairpin, whereas the Klenerman group does FRET measurements, which are sensitive not only to the transitions between open and closed states, but also to conformational fluctuations within the open state, which Libchaber's measurements would probably not detect. If conformational fluctuations within the open state are occurring on the same time scale as the opening and closing of the hairpin, the kinetics would deviate from single-exponential.

Marko and co-workers (Cocco et al., 2003a) have applied the kinetic zipper model for the opening and closing of an RNA hairpin that is held at a constant force of a few pN, to simulate the force-induced unfolding measurements of Bustamante and co-workers (Liphardt et al., 2001). They assume that the kinetic step corresponding to the opening of each base-pair is independent of force and is proportional to the exponential of the base-pairing free energy, while the closing of each base-pair is proportional to the exponential of $F\Delta x$, where F is the applied force, and Δx is the distance that has to be overcome against the applied tension to form the base-pair. The time-scale for each elementary step is set by a microscopic rate (r), which is a free parameter in their model. In order to describe the time-scale of ~1 second for the opening and closing time in the experiment of Bustamante and co-workers (Liphardt et al., 2001), the microscopic rate coefficient was found to be $r \sim 3.6\times10^{6}\ \mathrm{s}^{-1} \approx (280\ \mathrm{ns})^{-1}$ at 25 °C. Thus, in the absence of any force, this model suggests that the time required to close each base-pair is ~300 ns, which gives the closing time for a hairpin with ~10 bases in the stem of ~3 μs, independent of the sequence composition. Therefore, in the model of Cocco et al. (2003a), the slow closing times of hairpins is not from the slow formation of the looped conformations, but from the slow zipping of the stem by successive closing of base-pairs along the stem, one pair at a time. Their model makes a prediction that the closing times should scale linearly with the length of the stem. These predictions have yet to be tested in any systematic way for simple ssDNA and RNA hairpins. It is interesting to note that Grunwell et al. (2001) report a slight increase in the closing times, from ~133 ms to ~142 ms, when the stem size of their hairpin is increased from 7 to 9 base-pairs.

3.2. What is the activation enthalpy for the hairpin closing step?

Accurate measurements of the activation enthalpies and free energies associated with the transition state of a reaction step are critical for an understanding of the underlying mechanism. The activation enthalpy for the loop closure step, estimated from a range of thermodynamics and kinetics measurements, varies widely in magnitude and sign. Early thermodynamics studies of Uhlenbeck et al. (1973) report this number to be +25 kcal/mol for a $A_6C_NU_6$ hairpin, while Gralla and Crothers (1973) find it to be ~0 kcal/mol for a $A_4GC_5U_4$ hairpin. More recently, we applied the equilibrium zipper model, in which we calculated the free energy of each microstate in the ensemble, to describe the equilibrium melting profiles of ssDNA hairpins (Kuznetsov et al., 2001; Shen et al., 2001), and used the zipper parameters to reconstruct the free energy surface along an effective one-dimensional coordinate, defined as the fraction of intact base-pairs; see Figure 7. The transition state along this reaction coordinate was identified as an ensemble of looped conformations stabilized by one base-pair closing the loop. An Arrhenius-like plot, with $\Delta G^{\ddagger}/RT$ plotted versus inverse temperature, yields the enthalpy of the effective transition state to be about −9 kcal/mol relative to the unfolded state (inset to Figure 7).

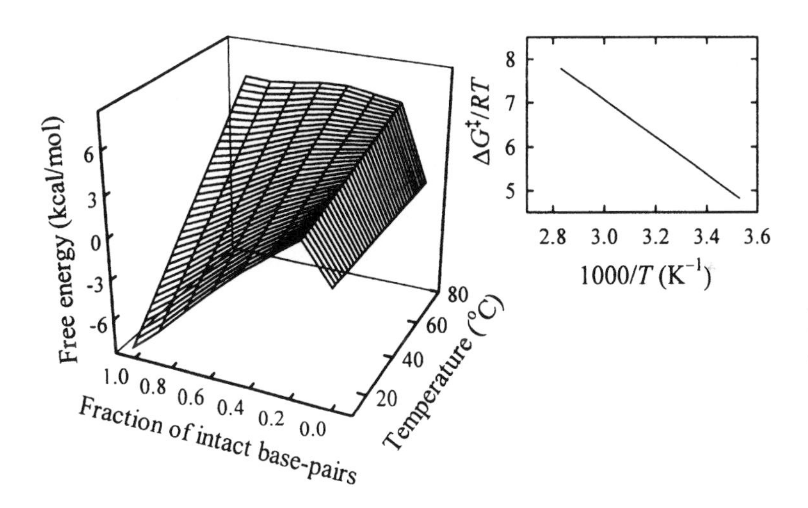

Figure 7. Free energy profiles versus the fraction of intact base-pairs and temperature for the hairpin 5'-CGGATAA(T_8)TTATCCG-3', obtained from the equilibrium zipper model of Kuznetsov et al. (2001). Inset; $\Delta G^{\ddagger}/RT$ is plotted versus inverse temperature. The values of $\Delta G^{\ddagger}$ are obtained from the free energy profiles. The slope on this plot gives $\Delta H_c^{\ddagger}/R$.

Kinetics data also gives widely varying values for the enthalpy of hairpin loop-closure. Libchaber's group reports positive activation enthalpies for the closing step for all hairpins in their study, with values ranging from +(5-15) kcal/mol for hairpins containing 8-30 bases in the loop, as illustrated in Figure 8 (Bonnet et al., 1998; Goddard et al., 2000). In contrast, our T-jump measurements yield negative activation enthalpies ~ –(10-13) kcal/mol for the closing step, in close agreement with the predictions from the equilibrium zipper model (Ansari et al., 2001; Kuznetsov et al., 2001). The Klenerman group reports non-Arrhenius dependence of the opening and closing times for their hairpin (Wallace et al., 2001).

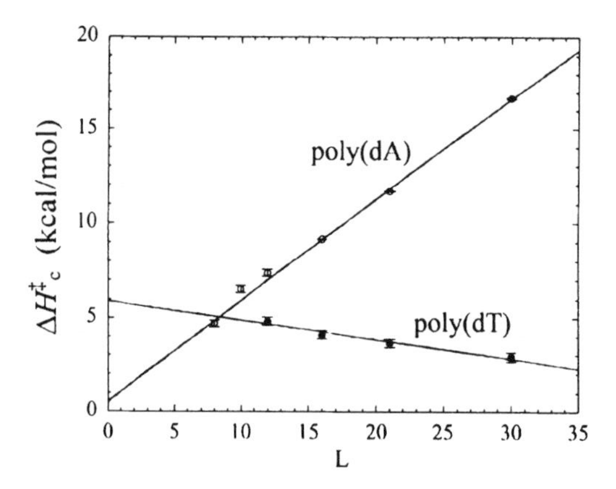

Figure 8. The activation enthalpy ($\Delta H_c^\ddagger$) for the closing step versus the number of bases (L) in the loop. The figure is adapted from Goddard et al. (2000). Open symbols are for hairpins with poly(dA) loops, and filled symbols are for hairpins with poly(dT) loops.

The comparison between the FCS measurements and the T-jump measurements can only be a qualitative one because there is, unfortunately, no overlap in the sequence of hairpins investigated by the different experimental groups. It is therefore not so clear whether the differences in the various sets of measurements are because of difference in sequences, or some inherent differences in the data acquisition and analysis. Earlier we discussed the differences in the two sets of FCS measurements. Here, we will point out two differences between the FCS and the T-jump measurements. One, in the FCS measurements, the equilibrium melting transitions as well as the kinetics are followed by monitoring the changes in the fluorescence emission of the fluorophore attached to one end of the hairpin sequence, and which loses its fluorescence intensity either from contact by a quencher attached to the other end, or from FRET, whereas in the T-jump measurements, the equilibrium profiles and kinetics are obtained from measurements of the changes in absorbance at 268nm. Whether the fluorescence changes, that monitor the dynamics of the ends of the hairpin, and absorbance changes that monitor the average property of all base-pairs, will yield identical melting profiles and kinetics has not really been established for these hairpins. It is possible that the fluorescence of tags attached at the ends would be most sensitive to the fraying of the hairpins at the ends, and which could be significantly different from the average absorbance measurements.

Another source of difference is in the temperature range over which the FCS and T-jump measurements are made. The FCS measurements of Libchaber's group are in the temperature range from an upper limit of ~50°C down to ~18°C (and in some instances down to ~10°C) for hairpins whose melting temperatures range from T_m = 30°C to 60°C for poly(dA) loops and T_m = 40°C to 60°C for the poly(dT) loops (Goddard et al., 2000). Thus, the bulk of their measurements are at or below T_m. The temperature range of the T-jump measurements is much narrower and hovers near T_m where the change in population as a result of T-jump is the largest. A possible explanation of the differences in the measured activation enthalpy for the closing step may be that the apparent activation enthalpy, obtained from slopes on Arrhenius plots, changes sign as the temperature is lowered below T_m, with negative activation enthalpies for $T \approx T_m$ and positive activation enthalpies for $T < T_m$. In fact, deviations from an Arrhenius behavior are evident even in the data from the Libchaber group; see Figure 9.

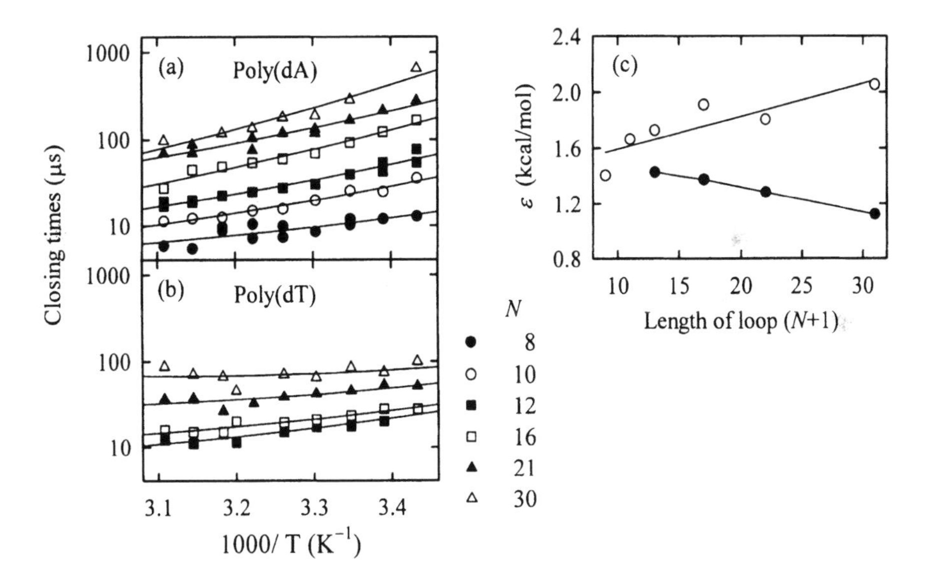

Figure 9. The closing times for (a) poly(dA) loops and (b) poly(dT) loops, with N bases in the loops, versus inverse temperature. The data are from Goddard et al. (2000). The continuous lines are fits to the data using the configurational diffusion model (Shen et al., 2001; Ansari et al., 2002). (c) The values of the characteristic roughness ε that describes the temperature dependence of the closing times for poly(dA) loops (open circles) and for poly(dT) loops (filled circles).

As discussed in Section 2.2, non-Arrhenius behavior could result from the temperature dependence of the enthalpy and entropy changes, or from the temperature dependence of the preexponential in Eq. (5). If the folding dynamics is modeled as configurational diffusion along the free energy profiles calculated from the equilibrium model, the non-Arrhenius behavior comes from the temperature dependence of the effective diffusion coefficient of the chain in the unfolded state (Ansari et al., 2001). In this model, the intrinsic diffusion coefficient of the ss-chain is modified by the factor $\exp[-(\varepsilon/RT)^2]$ where ε represents the roughness in the free energy surface between the unfolded state and the transition state as a result of transient intrachain interactions (Ansari et al., 2001). Since the diffusion coefficient appears in the preexponential for the closing step, the apparent activation enthalpy in an Arrhenius description has two contributions: one from the enthalpy of the effective transition state relative to the unfolded state (which is negative), and another from the temperature dependence of the diffusion coefficient (which is positive $\sim\varepsilon^2/RT$). Therefore, the closing times are expected to be small below the melting temperature as a result of deeper traps, and again small at high temperatures because of the intrinsic lower enthalpy of the effective transition state, leading to a non-Arrhenius temperature dependence. Applying the configurational diffusion model to the Libchaber data, with the assumption that intrachain interactions transiently trap the polynucleotide in misfolded conformations, reproduces the effective positive activation enthalpies of the Libchaber group, including the slight non-Arrhenius behavior observed in their data (Figure 9) (Shen et al., 2001; Ansari et al., 2002).

Non-Arrhenius temperature dependence for the closing times also comes out of the statistical mechanical model of Zhang and Chen (2002). Figure 10 shows a graph of the relaxation times versus inverse temperature for their RNA hairpin. For temperatures greater than the melting temperature, $T_m \approx 62$ °C for their hairpin (or $1/(k_B T_m) \approx 1.5$ mol/kcal), the relaxation times are dominated by the opening times, and yield positive activation enthalpy for the opening step, as expected. Below T_m, the relaxation times are dominated by the closing times and exhibit a distinctly non-Arrhenius temperature dependence, with a rollover at what they call the glass transition temperature (T_g), which for their hairpin is around 20°C ($1/(k_B T_g) \approx 1.7$ mol/kcal). They get negative activation enthalpy for the closing step between T_m and T_g, and positive activation enthalpy below T_g. This roll-over for $T < T_g$ is a consequence of misfolded states that behave as deep traps, so that, at these low temperatures, the rate-determining step for forming hairpins is to overcome these traps. Thus, their conclusions are in accord with the results of our configurational diffusion model on a rough energy surface.

3.3. Is a semiflexible polymer description of ss-polynucleotides valid?

Force-extension measurements that monitor the elastic response of a biopolymer have unambiguously demonstrated a semiflexible polymer description of double-stranded DNA (Bustamante et al., 1994; Marko and Siggia, 1994). It might be noted that this was

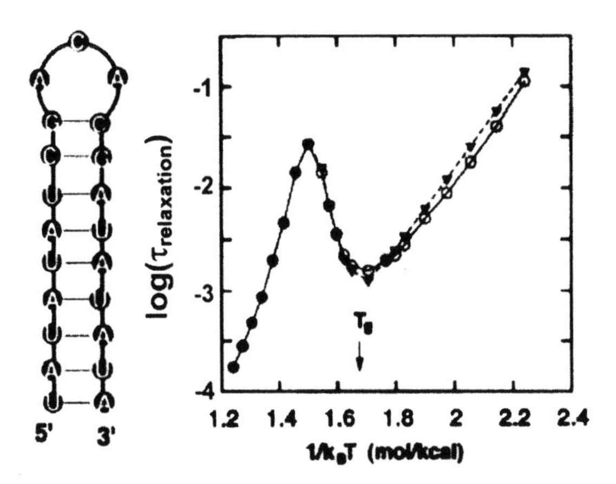

Figure 10. Relaxation times versus inverse temperature for an RNA hairpin. The figure is adapted from Zhang and Chen (2002). The relaxation times are from their statistical mechanical model for the hairpin shown on the left; the time-scale on the y-axis is in units of milliseconds (Shi-Jie Chen, private communication). T_g is the glass transition temperature.

previously quite solidly established by an array of different and consistent experimental approaches (Hagerman, 1988); however the force-extension measurements make the interpretation of DNA flexibility very clear. The theoretical description of ssDNA as a semiflexible polymer, on the other hand, requires several adjustments, at the least because of the non-negligible intrachain interactions (Figure 11). The first such measurements on ss λ-phage DNA (in 150 mM NaCl) supported a polymer description of a freely-jointed chain with a Kuhn's (statistical segment) length of ~1.5 nm (Smith et al., 1996). However, deviations from a semiflexible polymer description have been observed in the force-extension measurements for both high ionic conditions (e.g., 5 mM $MgCl_2$) and low ionic conditions (e.g., 2 mM NaCl), especially in the limit of low (< 10 pN) forces (Bustamante et al., 2000; Maier et al., 2000; Wuite et al., 2000). The low force behavior under high ionic solutions has been explained by various theoretical models as arising from secondary structures (hairpins) that form as a result of base-pairing interactions along a ssDNA. Thus, the forces required to initially stretch ssDNA have to overcome the base-pairing interactions and are found to be in excess of forces required to just overcome the entropic elasticity (Gerland et al., 2001; Montanari and Mezard, 2001; Zhang et al., 2001; Cocco et al., 2003b). The behavior at low ionic conditions has been explained as arising from the increased charge and hence the increased electrostatic repulsion of the ssDNA segments, resulting in an increase in the effective statistical segment length of the chain (Zhang et al., 2001; Cocco et al., 2003b).

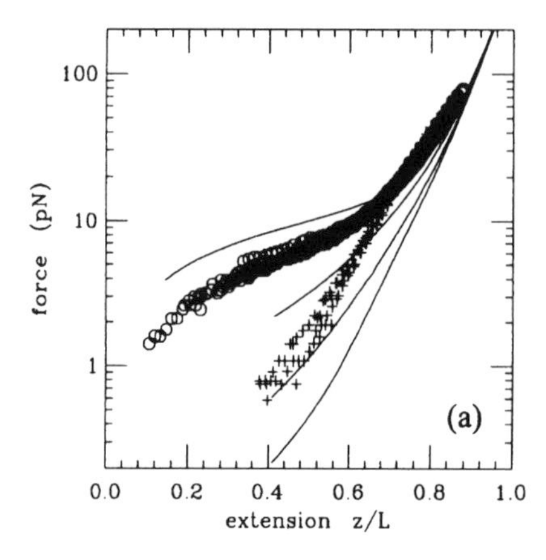

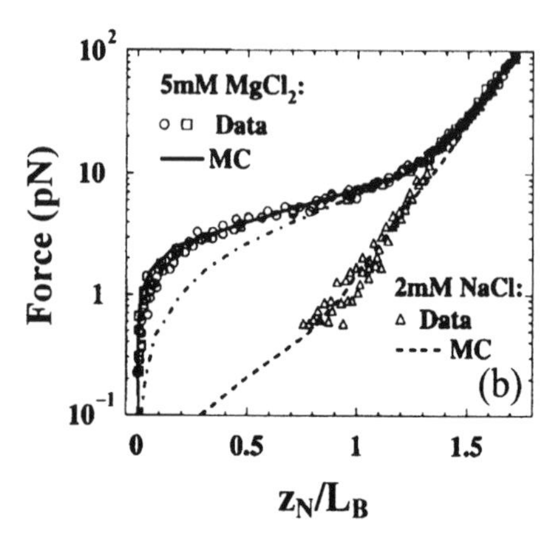

Figure 11. Force versus extension for ssDNA at various ionic strengths. (a) This figure is adapted from Cocco et al. (2003b); experimental data at 150 mM Na^+ (o) and 2.5 mM Na^+ (+) are from Bustamante et al. (2000). The continuous lines are from the theory of Cocco et al. (2003b) for, from top to bottom, 1.5 M, 150 mM, 15 mM, 1.5 mM, and 0.5 mM NaCl concentrations. (b) This figure is adapted from Zhang et al. (2001). The experimental data (Δ, □) are from Bustamante et al. (2000) and (o) are from Maier et al. (2000). The lines are from the calculations of Zhang et al. (2001) for 2 mM NaCl (dashed line) and 5 mM $MgCl_2$ (continuous line). The dash-dotted line is for a freely-jointed chain model without any interactions of the statistical segments.

Another measure of the semiflexible polymer nature of ssDNA comes from the dependence of loop-closure probability on the length of the loop. The simplest description of loop closure suggests that the closing time should scale as $\tau_c \sim L^2$, where L is the length of the loop, and that the stabilizing free energy of the hairpin should increase with decreasing loop size as $\sim 2RT\ln(L)$, for loop sizes $L \geq 10$ (Mathews et al., 1999). The dependence of the melting profiles of ssDNA hairpins on the loop size, however, show that the hairpin stability deviates quite significantly from that expected for an ideal polymer, and varies as $\sim \alpha RT\ln(L)$ where $\alpha \sim 7$ for loops ranging from 4-12 bases, and α ~3-5 for loops ranging from 10-30 bases (Kuznetsov et al., 2001; Shen et al., 2001). Therefore smaller loops are much more stable than expected from entropy considerations alone, presumably from favorable stacking interactions within the loop and exclusion of water in tighter loops (Vallone and Benight, 1999) (Figure 12)

The equilibrium measurements then raise the question, how do the opening and closing times scale with loop size? Libchaber and co-workers found that the opening times were insensitive to both the loop sequence and length whereas the closing times scaled as $\sim L^{2.6}$ for poly(dT) loops ranging in length from 12-30 bases. These observations are not so inconsistent with what is expected for a semiflexible polymer (Aalberts et al., 2003). Their closing times for poly(dA) loops exhibit a slightly stronger dependence of $\sim L^{3.2}$ (as estimated from their data, shown in Figure 9). The most striking result from

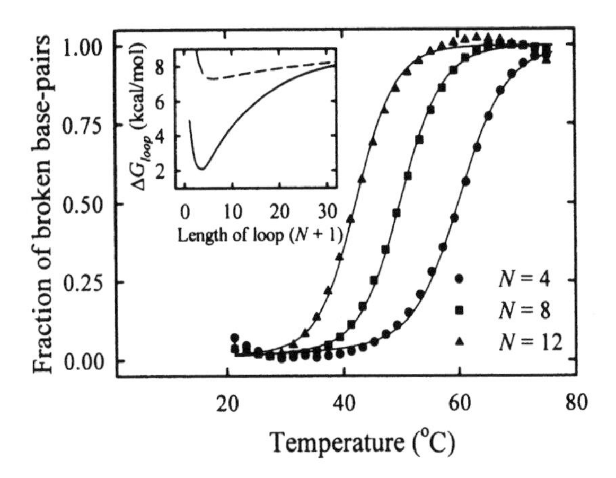

Figure 12. Melting profiles for hairpins with poly(dT) loops with N bases in the loop. The solid lines are a fit to the data with an equilibrium zipper model (Kuznetsov et al., 2001). Inset: The free energy of forming a loop closed by a single base-pair (ΔG_{loop}) versus the number of bases in the loop. The continuous line is a plot of the ΔG_{loop} values that fit the observed loop dependence of the melting profiles; the dashed line is a plot of the ΔG_{loop} values expected if stacking interactions within the loop are ignored.

Libchaber and co-workers, and which demonstrates qualitatively very different behavior for poly(dT) versus poly(dA) strands, is the loop-size dependence of the apparent activation enthalpy for the closing step (Figure 8). The activation enthalpy is found to be ~5 kcal/mol for both poly(dA) and poly(dT) loops ~10 bases long. For poly(dT) loops, they observe a slight decrease in the activation enthalpy with increasing loop size. However, for poly(dA) loops the apparent activation enthalpy increases quite dramatically, from ~5 kcal/mol to >15 kcal/mol for 30 bases long loops. Therefore, they find that the enthalpic barrier increases linearly with the number of bases in poly(dA) loops, with a slope of +0.5 kcal/mol/base. The authors conclude from this study that while the free energy of forming loops is mostly entropic for poly(dT) loops, the free energy of forming poly(dA) loops includes an additional enthalpic contribution, which arises from disrupting base stacking interactions in poly(dA) in order to form the loops. Based on the linear dependence of the apparent activation enthalpy on the length of the loop, they suggest that the number of stacking interactions that are disrupted increases linearly with the length of the loop, with ~0.5 kcal/mol of enthalpy cost for disrupting a single AA stacking interaction (Goddard et al., 2000).

It is well known that poly(dA) or poly(rA) form helical structure as a result of base-stacking and which give them a rigidity which is significantly larger than that of poly(dT) or poly(rU), especially at low temperatures (Eisenberg and Felsenfeld, 1967; Inners and Felsenfeld, 1970; Stannard and Felsenfeld, 1975). Therefore, for loops that are smaller than the persistence length of the strands, an enthalpic cost of deforming the chain to form a loop is to be expected (Goddard et al., 2002). However, whether the persistence length of poly(dA) chains is large compared to the length scales of 8-30 bases is not completely resolved, and is highly temperature dependent. As the temperature is raised, the ssDNA does start to show behavior reminiscent of a semiflexible polymer with dynamics that are not so strongly coupled to the sequence. The most compelling evidence of this is in the scaling of the closing times (τ_c) with the loop length (L) near the T_m of the hairpin. In T-jump measurements τ_c scales as $\sim L^2$ for both poly(dT) and poly(dA) loops in the range of 4-12 bases (Figure 13) (Shen et al., 2001; Ansari et al., 2002). The main point to note here is that in the vicinity of T_m, the hairpins show nearly identical behavior for both types of loops.

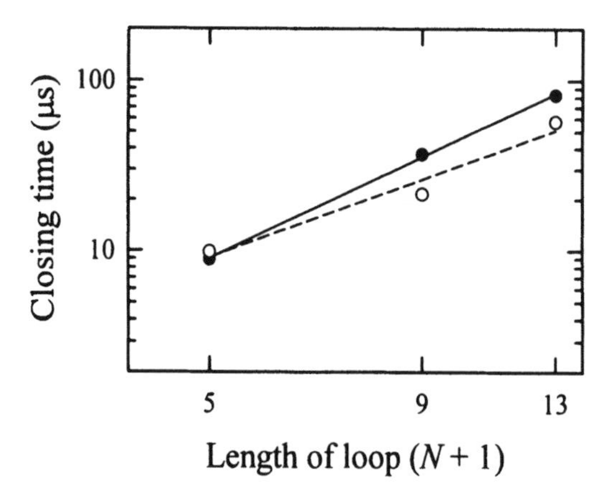

Figure 13. Closing time τ_c versus length of the loop from T-jump measurements. τ_c scales as $\sim L^{2.2}$ for poly(dT) loops (•) at 51°C and as $\sim L^{1.8}$ for poly(dA) loops (○) loops at 43°C.

As discussed in Section 2, slopes on Arrhenius plots yield activation enthalpies if the entropy and enthalpy changes as well as the preexponential factor are temperature independent. Since this is rarely the case, one ends up obtaining apparent activation enthalpies that need to be interpreted with caution. In our configurational diffusion model, the apparently anomalous values of activation enthalpies for the closing step as the temperature is lowered arise as a result of the temperature dependence of the preexponential in the Arrhenius expression. This model explains, albeit qualitatively, the observation that the activation enthalpy for poly(dA) loops increases with increasing loop length, since poly(dA) strands have a greater tendency to stack, or mis-stack, as the intervening chain length increases, thus increasing the roughness in the energy surface (Figure 9).

Here, we will describe an alternative model, proposed by Aalberts et al. (2003) that also captures the increase in the apparent activation enthalpy for the closing step with increasing loop size in poly(dA). Aalberts et al. retain the semiflexible polymer nature of ssDNA, but introduce a temperature dependence to the persistence length for poly(dA) strands but not for poly(dT) strands. They write down an analytic expression for the persistence length, in units of nucleotides, as

$$P = \frac{1}{\ln\left[1 + \exp\left(\Delta G_s / RT\right)\right]} \tag{10}$$

where $\Delta G_s = \Delta H_s - T\Delta S_s$ is the stacking free energy per stack for a ss-chain. They estimate the loop closing entropy as $\exp(\Delta S_{loop} / R) \sim r_c^3 / < r^2 >^{3/2}$, where r_c is the threshold distance at which contact is made between two bases closing the loop, and $<r^2>$ is defined in Eq. (9). Assuming only an entropic contribution to the closing rate yields $k_c \sim \exp(\Delta S / R)$. If the persistence length were temperature independent, the Aalberts models would yield temperature independent closing rates and zero activation enthalpy for the closing step. However, since the persistence length is temperature-dependent, an Arrhenius plot of the logarithm of the closing rate versus inverse temperature yields an apparent activation enthalpy $\Delta H_c^\ddagger$, which depends on the length of the loop via Eq. (9).

In order to estimate $\Delta H_c^\ddagger$, Aalberts and co-workers obtain estimates of ΔH_s and ΔS_s by fitting the temperature dependence of the experimentally measured closing times for poly(dA) and poly(dT) loops based on a model in which the closing times for poly(dA) loops depend on the number of stacked pairs (as the number of stacks increases, the closing rate decreases), with closing times for poly(dT) loops defined as the closing times for chains with no stacks. The temperature dependence of the closing times in poly(dA) loops enters via the Boltzmann factor $\exp(-n\Delta G_s/\mathrm{RT})$ that is used to define the probability of finding n stacks at a given temperature T. For a particular chain with n stacks, they simulate the closing times using a Monte-Carlo procedure, and vary the parameters ΔH_s and ΔS_s to obtain agreement with the experimental values of the closing times obtained by Libchaber and co-workers. They find that values of ΔH_s ranging from −4.5 kcal/mol to −8.2 kcal/mol and ΔS_s ranging from −14 cal/mol/K to −25 cal/mol/K give good agreement between experiment and simulations. Figure 14 shows the $\Delta H_c^\ddagger$ calculated from their model at $T = 310$K, using stacking parameters $\Delta H_s = -9$ kcal/mol and $\Delta S_s = -25.6$ cal/mol/K. The calculated values are in good agreement with the experimental values of $\Delta H_c^\ddagger$ obtained by Libchaber and co-workers. Note that their estimate of ΔH_s is significantly larger than the ~0.5 kcal/mol estimate of Libchaber and co-workers, and is in closer agreement with previous estimates of the stacking parameters (Turner, 2000).

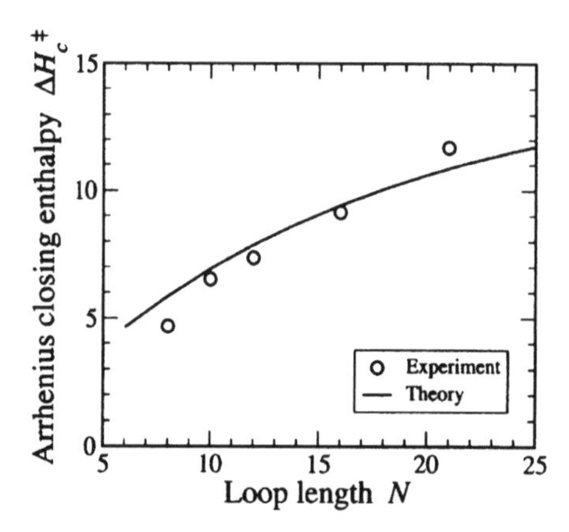

Figure 14. Apparent activation enthalpy for the closing step ($\Delta H_c^{\ddagger}$) as a function of the number of bases in the poly(dA) loop. Symbols are the experimental results of Goddard et al. (2000), also shown in Figure 8. The line is a fit to the data from the model of Aalberts et al. The figure is adapted from Aalberts et al. (2003).

There have been several attempts in the past to measure the persistence length of ss-polynucleotides, with values obtained from a number of different experiments ranging from 0.75–7.5 nm (Eisenberg and Felsenfeld, 1967; Inners and Felsenfeld, 1970; Smith et al., 1996; Tinland et al., 1997; Rivetti et al., 1998; Mills et al., 1999; Kuznetsov et al., 2001; Shen et al., 2001). Most of the measurements were done at a single temperature. The only systematic study of the temperature and sequence dependence of the persistence length of ss-polynucleotides was conducted by Felsenfeld and co-workers (Eisenberg and Felsenfeld, 1967; Inners and Felsenfeld, 1970). They showed that the radius of gyration, obtained from light scattering measurements, showed only a weak temperature dependence for poly(rU) chains in the range 15°C to 45°C, whereas the radius of gyration of poly(rA) chains increased by more than a factor of 2 from 40°C to 0°C. Both poly(rU) and poly(rA) chains of the same length had similar dimensions at temperatures above about 50°C, for which the poly(rA) chain is essentially unstacked. Using their measured values of the radius of gyration for a poly(rA) chain of length 1462 nucleotides at different temperatures, one can calculate the persistence length values as a function of temperature. These values are plotted in Figure 15, together with the corresponding values from Aalberts et al. (Eq. 10). The two sets of numbers agree reasonably well at high temperatures, but show deviations at low temperatures, with the numbers from Aalberts et al., showing a stronger temperature dependence. Thus, although the model and calculations of Aalberts et al. captures the anomalous dependence of the activation enthalpy for the closing step on the length of the poly(dA) loops, their estimates of the persistence lengths of poly(dA) loops at low temperatures are significantly larger than previous estimates.

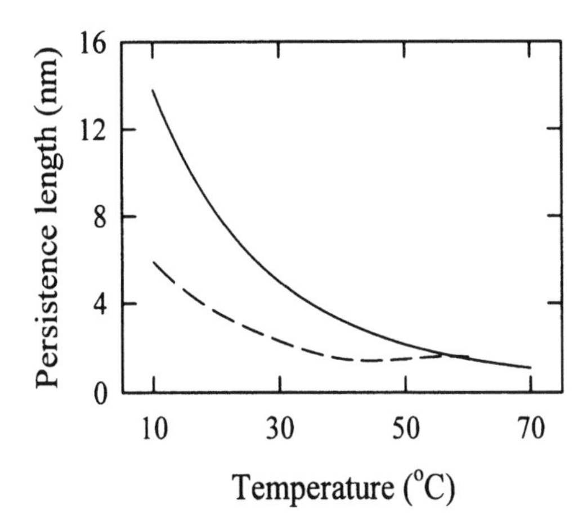

Figure 15. Persistence length (in nm) versus temperature. Top curve (continuous line) is from the model of Aalberts et al. (Eq. 10), with $\Delta H_s = -9$ kcal/mol and $\Delta S_s = -25.6$ cal/mol/K. The bottom curve (dashed line) is from the radius of gyration measurements of Eisenberg and Felsenfeld (1967) on poly(rA). An internucleotide distance of 0.6 nm was used.

The most straightforward experiment that might help to unravel some of these issues is force-extension measurements on ss-poly(dT) and ss-poly(dA) that will not have the complications of hairpin formation, as observed for ss λ-phage DNA, but may reveal differences in the stacking interactions that (a) just change the persistence length and (b) lead to mis-stacked clusters under low-force conditions.

3.4. What is the viscosity dependence of the opening and closing rates?

Studies on the viscosity dependence of the overall rates are important because they can provide new insights into the rate-determining processes, and also provide additional constraints on models that are used to describe the dynamics of hairpin formation. Klenerman and co-workers report that the opening and closing times scale with the solvent viscosity (η) as $\sim\eta^{0.8}$ (Wallace et al., 2001). One of the major obstacles in interpreting kinetics obtained from measurements in solvents of varying composition is that addition of viscogenic cosolvents invariably affects the stability of the hairpins. Separating the effects of cosolvents on the rate coefficients from changes in stability or from changes in the viscosity of the solvent is non-trivial. It is not straightforward to interpret the viscosity dependence of measured relaxation times without first correcting for these inevitable changes in stability. Wallace et al. assume that addition of nearly 50% glycerol does not affect the stability of their hairpins; they conclude that all the changes in the observed rates upon addition of glycerol can be attributed to changes in the solvent viscosity.

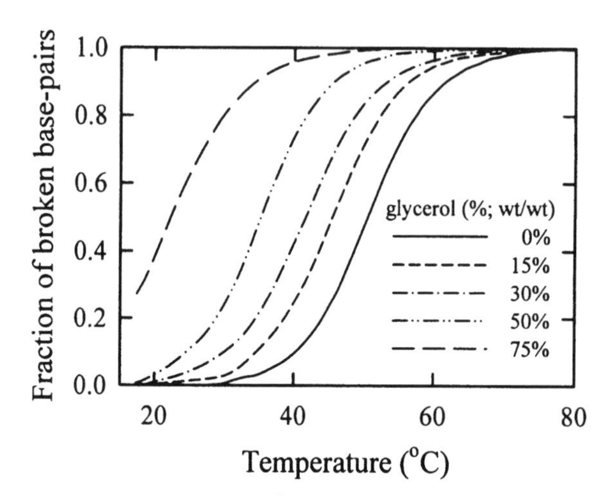

Figure 16. Melting profiles of hairpin 5'-CGGATAA(T_8)TTATCCG-3' in solutions of varying glycerol/water mixtures. Addition of glycerol changes the viscosity as well as the stability of the hairpin.

Early attempts of Wang and Davidson to measure the viscosity dependence of the DNA cyclization times showed that addition of glycerol significantly destabilizes the base-pairing (Wang and Davidson, 1968). We have measured the melting profiles of hairpins as a function of glycerol concentration and find that T_m decreases by more than 15°C upon addition of 50% (w/w) glycerol and by nearly 30°C upon addition of 75% (w/w) glycerol (Figure 16). One way to compensate for changes in stability is to assume that an increase in the free energy of the hairpin (relative to the unfolded state) upon addition of glycerol ($\Delta\Delta G$) is reflected in a corresponding increase in the free energy of the transition state ($\varepsilon\Delta\Delta G$, with ε between 0 and 1) (Figure 17). The hairpin closing (τ_c) and opening (τ_o) times can then be written as:

$$\begin{aligned}\tau_c(\eta) &= \tau_c(\eta_0)\left(\frac{\eta}{\eta_0}\right)^{\kappa} \exp\frac{\Delta G^{\ddagger}_{cw} + \varepsilon\Delta\Delta G}{RT} \\ \tau_o(\eta) &= \tau_o(\eta_0)\left(\frac{\eta}{\eta_0}\right)^{\kappa} \exp\frac{\Delta G^{\ddagger}_{ow} + (\varepsilon-1)\Delta\Delta G}{RT}\end{aligned} \quad (11)$$

Here $\Delta G^{\ddagger}_{cw}$ ($\Delta G^{\ddagger}_{ow}$) is the free energy barrier for the closing (opening) step in water, and κ is a parameter between 0 and 1 that describes the viscosity dependence (S.V.K & A.A., manuscript in preparation).

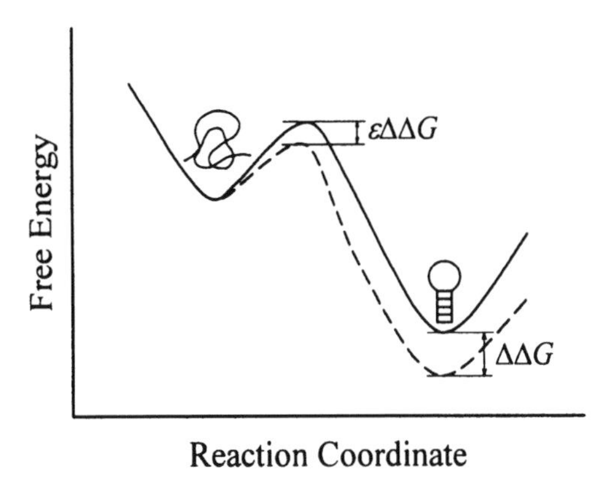

Figure 17. A schematic representation of the free energy versus an effective reaction coordinate for hairpin formation for two different concentrations of a viscogenic solvent. A linear free energy approximation is assumed in which the transition state is destabilized by some fractional amount ($\varepsilon\Delta\Delta G$) when the hairpin state is destabilized by an amount $\Delta\Delta G$.

Figure 18 shows the relaxation times for hairpin 5'-CGGATAA(T_8)TTATCCG-3' measured for different concentrations of glycerol in the sample. Figure 19 shows the viscosity dependence of the relaxation times, and the opening and closing times, at 35°C. The opening and closing times are fit using Eq. (11), with ε and κ as free parameters. In this analysis we find that the apparent viscosity dependence of τ_c and τ_o can be described equally well by assigning most of the dependence to a change in the stability of the transition state ($\varepsilon \approx 0.6$) with weak viscosity dependence ($\kappa \approx 0.4$) or by assigning nearly all dependence to a change in solvent viscosity ($\kappa \approx 1$) and a smaller change in the stability of the transition state ($\varepsilon \approx 0.3$). Therefore, the conclusion that the opening and closing times scale linearly with solvent viscosity is ambiguous. The addition of glycerol or other viscogenic agents that destabilize hairpins can have an additional effect. Since glycerol destabilizes hairpins, it is also expected to destabilize any misfolded conformations, and reduce the roughness in the free energy surface. Thus, addition of glycerol may have two effects that could compensate: a decrease in the *intrinsic* diffusion coefficient with increasing viscosity, and an *increase* in the *effective* diffusion coefficient for intrachain dynamics as a result of weaker intrachain interactions. Therefore, a careful investigation of the viscosity dependence requires isostability conditions, e.g., compensating for the changes in stability upon addition of viscogenic cosolvents by varying the salt concentrations. A similar approach has been successfully applied to monitor the viscosity dependence of the protein folding kinetics (Jacob et al., 1997; Plaxco and Baker, 1998).

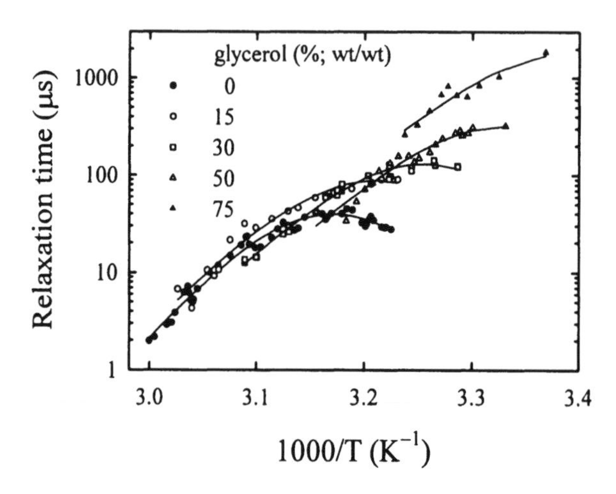

Figure 18. Relaxation times for the hairpin 5'-CGGATAA(T_8)TTATCCG-3' versus inverse temperature in solutions of varying glycerol/water mixtures.

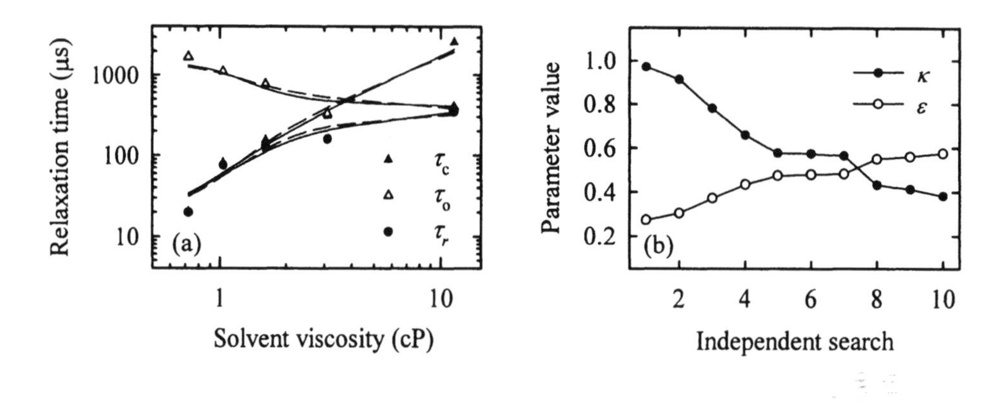

Figure 19. (a) Viscosity dependence of the relaxation times and the opening and closing times at $T = 35°C$. The continuous lines are a fit to the data using Eq. (11), with $\kappa \approx 1$ and $\varepsilon \approx 0.3$; the dashed lines are with $\kappa \approx 0.4$ and $\varepsilon \approx 0.6$. (b) The parameter values for 10 independent searches in parameter space that give equally good fit to the observed viscosity dependence.

3.5. Does transient trapping in misfolded states slow down hairpin formation?

In order to address the question of whether the rate-determining step in hairpin formation is transient trapping in misfolded states, we used a kinetic zipper model to simulate the relaxation kinetics (S.V.K & A.A., manuscript in preparation). This kinetic model is a simple extension of the equilibrium zipper model that we used previously to describe the melting profiles of ssDNA hairpins with loops of different sizes (Kuznetsov et al., 2001), with a simple modification in that we include explicitly all misfolded states with non-native base-pairs, as in the Zhang and Chen model (2002), with only the constraint that the loop cannot be smaller than 3 bases. The model, at present, does not include non-native hydrogen bonds and intrastrand stacking contacts between any two bases along the ss-chain, as has been observed in the molecular dynamics simulations of Pande and co-workers (Sorin et al., 2003), although the framework of the algorithm allows such interactions to be included without much difficulty.

In our equilibrium zipper model, we described the end-loop weighting function $w_{loop}(n)$ for a loop of n bases as (Kuznetsov et al., 2001).

$$w_{loop}(n) = \left(\frac{3}{2\pi b^2}\right)^{3/2} V_r g(n) \sigma_{loop}(n) \tag{12}$$

Here b =2P is the statistical segment length, V_r is a characteristic reaction volume within which the bases at the two ends of the loop can form hydrogen bonds, $g(n)$ is the loop-closure probability for a semiflexible polymer of length n monomers, and $\sigma_{loop}(n)$ describes the increase in the stability of the hairpin as a result of stabilizing interactions within the loop. $\sigma_{loop}(n)$ was parameterized as

$$\sigma_{loop}(n) = \langle\sigma\rangle^{1/2} + \frac{C_{loop}}{N_b^{\gamma}} \tag{13}$$

with C_{loop} and γ as fitting parameters needed to describe the strength of the stabilizing interactions in the loops and its dependence on loop size and N_b is the number of statistical segments in the loop. For large loops the value of $\sigma_{loop}(n)$ approaches the cooperativity parameter $\langle\sigma\rangle^{1/2} = (4.5\times10^{-5})^{1/2}$ (Wartell and Benight, 1985). Recall that the primary result from the analysis of the equilibrium melting profiles was that the interactions that stabilize loops favored the formation of smaller loops by a factor that was significantly large than that expected from entropic considerations alone (Figure 12: inset). A value of $\gamma\sim7$ in Eq. (13) was necessary to describe the dependence of the equilibrium melting profiles on the size of the loops for hairpins 5'-CGGATAA(X_N)TTATCCG-3', for loops consisting of both X=A and X=T, and for the number of bases in the loop N = 4, 8, or 12 (Kuznetsov et al., 2001; Shen et al., 2001).

The first step in describing the relaxation kinetics is to write down the set of coupled differential equations that describe the transitions between the various microstates in the ensemble (Munoz and Eaton, 1999; Zhang and Chen, 2002; Cocco et al., 2003a)

$$\frac{dp_i}{dt} = \sum_{j \neq i} \left[k_{j \to i} p_j - k_{i \to j} p_i \right] \tag{14}$$

where p_i is the population of the ith microstate ($i = 1, \ldots, m$) and $k_{j \to i}$ and $k_{i \to j}$ are the rates for transitions from state j to state i and from state i to state j, respectively. The matrix form of Eq. (14) (master equation) is

$$\frac{d\mathbf{P}}{dt} = \mathbf{M} \cdot \mathbf{P} \tag{15}$$

where $\mathbf{P}$ is a column vector col($p_1, \ldots, p_m$), and $\mathbf{M}$ is a $m \times m$ rate matrix with $M_{ij} = k_{j \to i}$, $i \neq j$ as the off-diagonal elements and $M_{ii} = -\sum_{j \neq i} k_{i \to j}$ as the diagonal elements. The time-dependent solution of the rate equations yields the change in population as a function of time $\mathbf{P}(t)$, and is obtained by diagonalizing the matrix $\mathbf{M}$ to obtain its eigenvalues (λ_i) and eigenvectors (U_i). The solution to Eq. (15) can be written as:

$$\mathbf{P}(t) = \exp(\mathbf{M}t)\mathbf{P}(0) = \mathbf{U}\exp(\lambda t)\mathbf{U}^{-1}\mathbf{P}(0) \tag{16}$$

where $\exp(\lambda t)$ is a $m \times m$ diagonal matrix with $\exp(\lambda_i t)$ as the diagonal matrix elements, $\mathbf{U}$ is a $m \times m$ matrix consisting of the eigenvectors, and $\mathbf{P}(0)$ is the column vector representing the populations of all microstates at $t = 0$.

In our T-jump experiments we monitor the change in transient absorbance at 266 nm, which is interpreted as a change in the fraction of intact base-pairs $\theta_f(t)$. To compare the relaxation kinetics from the model with the transient absorbance measurements, we calculate $\theta_f(t)$ from the calculated populations of the microstates as

$$\theta_f(t) = \frac{1}{N_s} \sum_{i=1}^{m} \left[b_i p_i(t) \right] \tag{17}$$

where N_s is the number of base-pairs forming the stem of the native hairpin, and b_i is the number of intact base-pairs in the ith microstate.

To describe the elementary rates of forming or breaking base-pairs, we assume that the first base-pair that closes a loop of length L has a closing rate given by $k_{loop}^{c} = D_{loop}^{0} \left(L_0 / L \right)^2$, where D_{loop}^{0} is the rate for forming loops of reference length L_0

consisting of 10 bases. The rate for loop formation, therefore, scales with loop length as $\sim 1/L^2$. The reverse rate, for opening the loops, is $k_{loop}^{o} = k_{loop}^{c} \exp(\Delta G_{loop}/RT)$, where ΔG_{loop} is the difference in free energy between the fully unfolded state and the microstate with a single intact base-pair. The values ΔG_{loop} are determined from the parameters that describe the melting profiles (see Figure 12). Similarly, the rate for adding a base-pair to an adjacent base-pair is given by k_{bp}^{+}, which we assume to be independent of temperature and sequence. The reverse rate is sequence dependent and is determined from $k_{bp}^{-} = k_{bp}^{+} \exp(\Delta G_{bp}/RT)$, where ΔG_{bp} is the difference in free energy between two microstates that are connected by the formation of a single contiguous base-pair. Thus, all rate coefficients in Eq. (14) can be calculated in terms of two parameters, D_{loop}^{0} and k_{bp}^{+}, and the equilibrium free energies of each of the microstates in the ensemble. As in the model of Cocco et al. (2003a), all the sequence dependence is in the opening rates.

We estimated the values of the two parameters in the kinetic zipper model by simulating the kinetics for the hairpin 5'-CGGATAA(T_8)TTATCCG-3', for which we have previously obtained the statistical weights of each of the microstates from fitting the equilibrium melting profiles to the equilibrium zipper model (Shen et al., 2001). This hairpin can adopt 28 conformations with native contacts and 40 conformations with non-native contacts (with a minimal loop size of 3 bases). Thus the total number of microstates for this hairpin is $m = 69$, including the unfolded state. The simulations yield two sets of parameters that give reasonable agreement with the measured relaxation time of ~18 μs for a T-jump from 42°C to 51°C (see Figure 4).

The first set of parameters yields $k_{bp}^{+} \approx 4.8\times10^{7}\ \mathrm{s}^{-1} \approx (20\ \mathrm{ns})^{-1}$, with the relaxation rates not sensitive to the value of D_{loop}^{0} from $10^{9}\ \mathrm{s}^{-1}$ to $10^{6}\ \mathrm{s}^{-1}$. In this limit, the rate-determining step for forming hairpins is the addition of the second base-pair to the equilibrium population of the looped conformations, and hence is not sensitive to the time-scale of formation of these looped conformations. The closing rate in this limit is given by $k_c \approx (k_{loop}^{c}/k_{loop}^{o})k_{bp}^{+}$ where the term in the parenthesis is the equilibrium population of the looped conformations relative to the unfolded state, and is given by $\exp(-\Delta G_{loop}/RT)$. Using a value of $\Delta G_{loop} \approx 4.4$ kcal/mol at 51°C yields $k_{loop}^{c}/k_{loop}^{o} \approx 1 \times 10^{-3}$, and $k_c \approx 5\times10^{4}\ \mathrm{s}^{-1} \approx (20\ \mu\mathrm{s})^{-1}$, which is close to $k_c \approx 3\times10^{4}\ \mathrm{s}^{-1} \approx (33\ \mu\mathrm{s})^{-1}$ obtained from solving the complete rate equations. The value of $k_{bp}^{+} \approx (20\ \mathrm{ns})^{-1}$ estimated by our model should be compared with $\sim(125\ \mathrm{ns})^{-1}$ estimated by Porscke (1974a), $\sim(300\ \mathrm{ns})^{-1}$ estimated by Cocco et al. (2003a), and $\sim(12\ \mathrm{ns})^{-1}$ and $\sim(780\ \mathrm{ns})^{-1}$ for A·U and C·G base-pairs, respectively, estimated by Zhang and Chen (2002).

The second set of parameters that yields relaxation rates consistent with the experimentally measured values is $D_{loop}^{0} \approx 1\times10^{4}\ \mathrm{s}^{-1} \approx (100\ \mu\mathrm{s})^{-1}$ and $k_{bp}^{+} \approx 2.8\times10^{8}\ \mathrm{s}^{-1} \approx$

$(4\ \text{ns})^{-1}$. This solution suggests a very slow rate for the formation of the first base-pair, occurring on tens of microseconds for loops of about 4 bases, followed by the rapid zipping of the stem. This slow step could arise from either intrachain interactions, which would slow down the effective diffusion coefficient for loop formation, or from an additional barrier for the chemical step of base-pair formation even when the two ends of the loop are within reaction distance, and which is not included explicitly in our model.

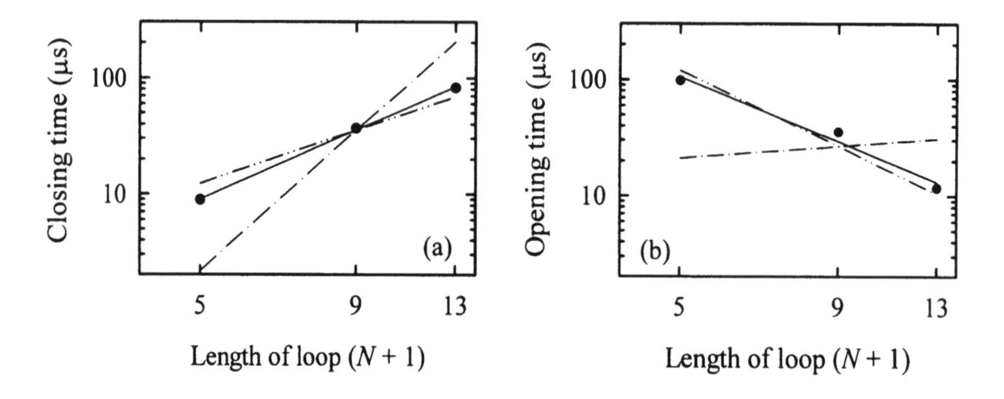

Figure 20. The closing times (a) and the opening times (b) versus the length of the loop for the hairpin 5'-CGGATAA(T_N)TTATCCG-3'; (•): experimental results from T-jump measurements; (—): a linear fit to the data with a slope of 2.2 in panel (a) and −2.3 in panel (b); (– · –) results of the kinetic zipper model for D^0_{loop} = $(40\ \text{ns})^{-1}$, k^+_{bp} = $(20\ \text{ns})^{-1}$, and (– ·· –) for D^0_{loop} = $(100\ \mu\text{s})^{-1}$, k^+_{bp} = $(4\ \text{ns})^{-1}$.

Next, we simulated the dependence of the relaxation times and the closing times on the size of the loop for the two sets of parameters. The first set yields a loop dependence for the closing times as $\tau_c \sim L^5$ (Figure 20), which is a much stronger dependence than what is observed experimentally (Bonnet et al., 1998; Shen et al., 2001). The reason that this set of parameters yields a strong dependence on loop size is because the loop dependence appears in the equilibrium constant $k^c_{loop}/k^o_{loop} = \exp(-\Delta G_{loop}/RT)$, and which was found to exhibit a much stronger dependence on loop-size than that expected for a semiflexible polymer (see Figure 12). The second set of parameters yields a loop dependence $\tau_c \sim L^2$, which is very close to the experimental results. This is to be expected from our model, since, in this limit $k_c \approx k^c_{loop}$, and we assume that $k^c_{loop} \sim 1/L^2$. Note that, in the T-jump measurements, the opening times are found to scale with loop size as $\sim L^{-2.3}$, which is also reproduced nicely by the second set of parameters.

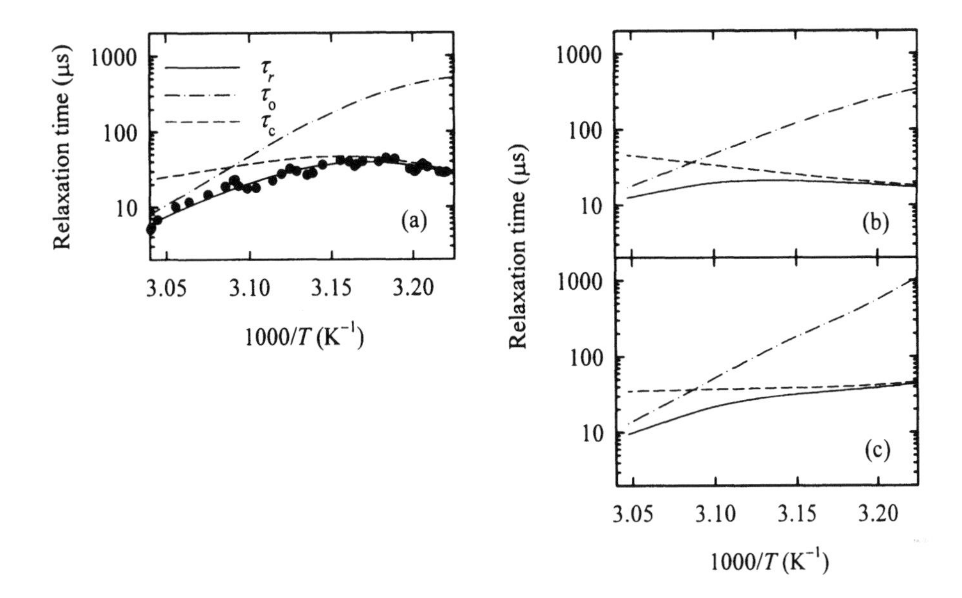

Figure 21. The relaxation times, closing times and opening times versus inverse temperature for the hairpin 5'-CGGATAA(T_8)TTATCCG-3'. (a) Symbols are relaxation times (τ_r) from T-jump experiments; the opening times (τ_o) and closing times(τ_c) at each temperature are from a two-state analysis of the measured relaxation times. Results from the kinetic zipper model are plotted in (b) for D^0_{loop} = (40 ns)$^{-1}$, k^+_{bp} = (20 ns)$^{-1}$, and in (c) for D^0_{loop} = (100 μs)$^{-1}$, k^+_{bp} = (4 ns)$^{-1}$. The line-types in panels (b) and (c) have the same meaning as in panel (a).

A further test of the parameters comes from simulating the temperature dependence of the relaxation times and the closing times for the hairpin 5'-CGGATAA(T_8)TTATCCG-3'. The results are shown in Figure 21 together with the corresponding experimental results. Both sets of parameters reproduce reasonably well the temperature dependence of the measured relaxation times. The primary difference between the two sets of parameters is the apparent activation enthalpy for the closing step, which is negative (~ – 2.3 kcal/mol) for the first set of parameters and essentially zero for the second set of parameters. The T-jump experiments yield negative activation enthalpies for the closing step near T_m. The two sets of parameters correspond to quite different physical pictures as described earlier. For the first set, $D^0_{loop} \approx (40\text{ ns})^{-1}$ and $k^+_{bp} \approx (20\text{ ns})^{-1}$, the rate-determining step is the formation of the second base-pair, and, since the rate for addition of base-pairs is assumed to be independent of temperature, the temperature dependence of the closing step is derived from the equilibrium population of the looped conformations, which increases with decreasing temperature, thus giving rise to an apparent negative activation enthalpy for the overall closing step in a two-state analysis. For the second set of parameters, D_{loop}^{0} = (100 μs)$^{-1}$ and k_{bp}^{+} = (4 ns)$^{-1}$, the rate-determining step is the formation of the first base-pair, which, in the current version of our model, is limited by chain dynamics and has little or no temperature dependence. Even if we were to assign an activation enthalpy to the chemical step of base-pair formation, it would be a positive activation enthalpy, and would not reproduce the experimental results.

The results of our simulations therefore present a puzzle in that the set of parameters that correctly describe the loop dependence are not able to describe the negative activation enthalpy for the closing times that is observed experimentally. The simulations and experimental results can be reconciled for the first set of parameters if we assume that the rate-determining step is indeed the formation of the second base-pair, but that in the effective transition state, which consists of looped conformations with one base-pair formed, all the loop stacking interactions that stabilize smaller loops (parameterized by C_{loop} in Eq. (13)) are not yet formed, so that the loop dependence of the equilibrium population of this transition state is still governed by the loss of conformational entropy of the chain, and which will give the correct scaling of closing time with the loop size. An alternative scenario, consistent with the simulations of Pande and co-workers, is that the collapse of the chain, followed by reorganization within the collapsed state, is the rate-determining step. Whether such a model would reproduce the scaling of closing times as ~L^2 remains to be investigated.

To investigate the effect of misfolded states on the kinetics of hairpin formation, we picked the following two hairpin sequences: (i) 5'-GGATAA$(T)_4$TTATCC-3' whose kinetics we have investigated previously (Ansari et al., 2001; Kuznetsov et al., 2001), and (ii) 5'-$A_6(T)_4T_6$-3', which has roughly the same stability as the first hairpin. Both hairpins have 21 conformations in the zipper model, not counting the misfolded states. The second hairpin can adopt many more misfolded conformations with non-native base-pair(s), 122 such conformations versus only 20 misfolded conformations for the first hairpin. Thus, the second hairpin is expected to have deep traps in its free energy surface, at temperatures below T_m, especially since some of the misfolded conformations (with 4 or 5 base-pairs in the stem) are nearly as stable as the fully intact hairpin.

The relaxation times and the closing times for these two hairpins were calculated as a function of temperature, using the same two sets of parameters as determined for the hairpin with T_8 bases in the loop, with one adjustment. For these hairpins, with T_4 bases in the loop, the k_{bp}^{+} parameter in the first set of parameters needed to be adjusted to ~5.5×10^7 s^{-1} $\approx$(182 ns)$^{-1}$ in order to yield relaxation times that agree well with the experimental value of ~13 μs measured for the hairpin 5'-GGATAA$(T)_4$TTATCC-3' at 37°C (Ansari et al., 2001; Kuznetsov et al., 2001). Therefore, for hairpins with T_4 bases in the loop, the k_{bp}^{+} parameter is approximately a factor of 10 smaller than the value of k_{bp}^{+} = (20 ns)$^{-1}$ estimated for the T_8 hairpin. This adjustment in k_{bp}^{+} is necessary for the first set of parameters that do not reproduce the dependence of closing times on the loop size correctly.

The relaxation kinetics was simulated for the folding of the hairpin when the temperature is dropped from high (~90°C), when the hairpin is completely unfolded, to the final temperature of interest. Although the T-jump in the simulations is reverse of the T-jump in experiments, we chose to go from high to low temperatures in order to reduce the amplitude of a fast phase that arises from the rapid zipping/unzipping of the stem, and which is observed in simulations of T-jumps from low to high temperatures, as well as in T-jump experiments on sub-microsecond time-scales (Porschke, 1974a). The folding kinetics of the hairpin with few (and not very stable) misfolded states are monophasic and well described by a single-exponential decay at all final temperatures ranging from 10°C to 95°C. For the second hairpin with many misfolded states, the kinetics observed for $T > T_m$ are also well-described by a single-exponential decay, with relaxation times that are about a factor of 6 smaller than that for the first hairpin. This apparently counter-intuitive result is a consequence of the fact that for $T > T_m$, the fully intact hairpin and some partially folded and misfolded hairpins are populated with similar probabilities at equilibrium, and the folding for the second hairpin occurs via many parallel pathways corresponding to the different conformations accessible in the "folded" state, thus speeding up the overall folding rate.

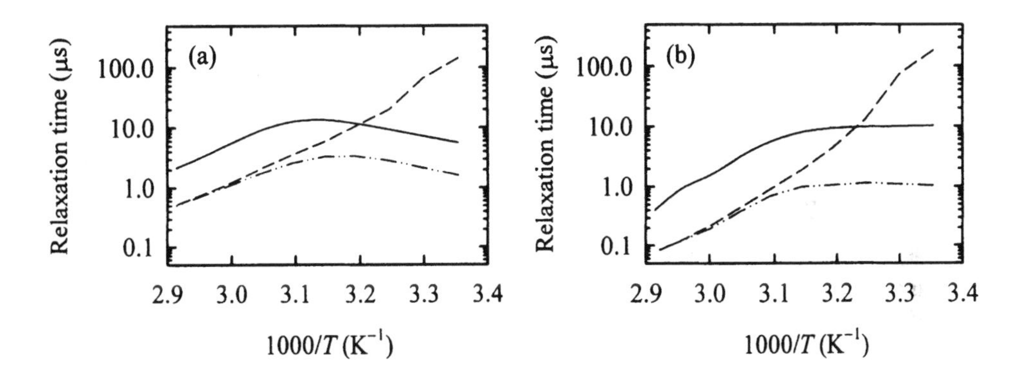

Figure 22. Relaxation times versus inverse temperature after T-jump from 90°C to 25°C, obtained from the kinetic zipper model for two hairpins; 5'-GGATAA(T_4)TTATCC-3' (with few misfolded conformations) and 5'-A_6(T_4)T_6-3' (with many misfolded conformations). Calculated relaxation times for (a) the parameters D^0_{loop} = (40 ns)$^{-1}$, k^+_{bp} = (187 ns)$^{-1}$, and for (b) the parameters D^0_{loop} = (100 μs)$^{-1}$, k^+_{bp} = (4 ns)$^{-1}$. The continuous line is for the first hairpin, and the dashed lines for the second hairpin; at low temperatures, the second hairpin exhibits two components in its relaxation kinetics.

For $T < T_m$, the folding kinetics of the second hairpin are distinctly biphasic, with a fast phase that corresponds, as before, to the rapid formation of both native and misfolded hairpins via multiple pathways, and then a slower reorganization of the misfolded conformations to form the correct native state. Figure 22 shows the calculated relaxation times as a function of temperature for the two hairpins. The results are qualitatively the same for both sets of parameters, with the hairpin with many misfolded states folding about 10 times slower than the hairpin with few misfolded states.

Finally, we also investigated the dependence of the relaxation times and the closing times on the length of the stem. For this simulation, we picked the hairpin sequences 5'-$A_6(T)_4T_6$-3' and 5'-$A_{24}(T)_4T_{24}$-3', and ignored the misfolded conformations for simplicity. Calculations were performed for T-jumps from 95°C to 25°C. The parameters we picked for this simulation were D^0_{loop} = (40 ns)$^{-1}$ and k^+_{bp} = (300 ns)$^{-1}$. These parameters are close to the parameters we found for our T_4 hairpins, and the rate for the addition of a base-pair is also very close to the corresponding parameter in the model of Cocco et al. (2003a). The results of our simulations yield k_c ≈(13 μs)$^{-1}$ for the 6-stem hairpin and k_c ≈(9 μs)$^{-1}$ for the 24-stem hairpin; i.e. the closing times for the two hairpins are nearly identical in this model, with the longer hairpin exhibiting slighter *faster* closing rates. For this set of parameters, the rate-determining step is the formation of the second base-pair, which depends on the equilibrium population of the looped conformations, and is independent, to first approximation, on the stem-length. The slightly faster rate for the longer hairpin is presumably from the larger number of parallel pathways arising from looped conformations of different sizes that the longer hairpin can adopt. Therefore, in contrast to the model of Cocco et al., our model suggests that even with a base-pair formation time as slow as 300 ns, longer hairpins do not fold any slower. Adding back the misfolded conformations is expected to slow down the longer stem hairpin at low temperatures because of the many more misfolding possibilities.

In summary, the simulations using the kinetic zipper model demonstrate that the relaxation kinetics of small hairpins can be described quite well over a range of temperatures and loop sizes in terms of a simple model with just two parameters. However, the results from the simulations reveal two sets of plausible parameters that reproduce different facets of the experimental data, but with very different rate parameters for the fundamental step of loop closure. The model also demonstrates that misfolded states can slow down hairpin formation, especially for $T < T_m$.

4. CONCLUSION

In this chapter, we have reviewed a vast array of experimental and theoretical studies of a fundamental problem in biopolymer science, the seemingly simple problem of the folding dynamics of hairpin structures in ss-polynucleotides. The review highlights the fact that, despite more than thirty years of biophysical research on this problem, some basic issues remain unresolved. First, the rate-determining step that defines the folding times of several tens of microseconds for short ssDNA or RNA hairpins in solution is not completely understood. Second, the scaling of the opening/closing rates with solvent viscosity has not been established.

The most fundamental question raised in this chapter may be succinctly stated as: Why do hairpins in ss-polynucleotides form so slowly? This question is opposite to the one posed in connection with the folding of proteins by Levinthal (1969), frequently referred to as the Levinthal paradox, where he argued that it should take longer than the age of the universe for proteins to fold if they randomly sampled all conformational space. Thus, the question underlying the field of protein folding became: Why do proteins fold so fast? (Baldwin, 1996). This paradox is resolved by recognizing that the conformational search for the native state of the protein is not random but biased, on a "funnel-like" energy landscape with a pronounced global minimum corresponding to the native state (Sali et al., 1994a; 1994b; Wolynes et al., 1995; Dill and Chan, 1997; Onuchic et al., 1997). Zwanzig et al. (1992) showed that if an energy penalty of the order of a few $k_B T$ is imposed to break favorable interactions that correspond to native interactions of the protein, then the conformational search time can be reduced to biologically relevant times.

In the case of hairpin formation in short ss-oligonucleotides, if intrachain interactions are ignored and the ss-chain treated like an ideal semiflexible polymer, the purely entropic conformational search time for the two ends of the chain to come together is tens of nanoseconds. However, force-extension measurements on ss-polynucleotides have demonstrated that, under conditions of high ionic strength, intrachain interactions such as random base-pairing and intrastrand stacking interactions are non-negligible, and lead to significant deviations from the behavior expected for a semiflexible polymer. Thus, one may ask whether the slow times observed for the folding of hairpins are because these intrachain interactions in the unfolded state lead to a roughness in the energy landscape that impedes the chain dynamics prior to the formation of the critical nucleus.

Several experiments suggest that sequence-dependent intrachain interactions may play a role in defining the rate-determining step. A comparison between the folding/unfolding dynamics of hairpins with identical stems but with different loop compositions, poly(dA) or poly(dT), have revealed many differences at temperatures below T_m: sequences with poly(dA) loops (i) take longer to fold than sequences with poly(dT) loops; (ii) exhibit apparent activation enthalpies for the closing step that increase with increasing loop size, in contrast with the behavior for poly(dT) loops; and (iii) exhibit deviations from the scaling of closing times as $\sim L^2$. The increased propensity of the poly(dA) loops to stack and to mis-stack may, to a large extent, be responsible for these differences. Experimental measurements of the viscosity dependence of the opening/closing times would likely lead to new insights into the rate-determining step. One trivial effect of the addition of viscogenic solvents is that, since these co-solvents destabilize base-pairing and stacking interactions, the differences in the behavior observed for the two kinds of loop strands should diminish. More importantly, an accurate determination of the scaling behavior with increasing solvent viscosity would provide additional constraints on the models. For example, significant deviations from a simple linear scaling could suggest that the conformational search for the critical nucleus occurs in the collapsed state, with internal rearrangement dominating over diffusion through the solvent.

An alternative scenario to explain the folding times of tens of microseconds for these hairpins is the slow and sequential zipping of the stem. How much of the slow folding times arise from conformational searching with transient traps in the free energy surface, and how much from the slow zipping of the stem, remains to be investigated in detail. There are currently no systematic studies on the dependence of hairpin closing times on stem length, analogous to the studies on loop-size dependence. Increasing the length of the stem will, of course, also increase the number of transient intrachain interactions, which would need to be included in a more complete theoretical description.

For hairpins especially designed to have a large number of misfolded states that act as deep traps, the numerical results of the kinetic zipper model show that hairpin sequences with many misfolded states exhibit an increase in folding times for $T < T_m$. Currently, this model ignores all interactions that are not base-pairing interactions. Including non-native hydrogen bonding and intrachain mis-stacking interactions would be a very useful extension to the kinetic zipper model, especially in light of the simulation results of Pande and co-workers (Sorin et al., 2003).

Finally, an essential experiment that will directly address the influence of intrachain interactions on the dynamics of the ss-polynucleotides is the measurement of the end-to-end contact time as a function of sequence composition, solvent conditions, and temperature. Such measurements, together with the insights on the semiflexible nature of the ss-polynucleotides gained from force-extension measurements, will be necessary to sort out the various contributions and provide a satisfying answer to this decades old folding problem.

5. ACKNOWLEDGEMENTS

We are indebted to John Marko for numerous discussions that we have benefited from and for his careful reading of the manuscript. We thank Shi-Jie Chen for reading an early draft of the manuscript, and Jay Joshi for help with some of the calculations. This work was supported by the donors of the Petroleum Research Fund through grant ACS-PRF 32099-AC4, and by the National Science Foundation through grants MCB-9722295 and MCB-0211254.

6. REFERENCES

Aalberts, D. P., Parman, J. M., and Goddard, N. L., 2003, Single-strand stacking free energy from DNA beacon kinetics., *Biophys. J.* **84**:3212.

Alberts, B., Bray, D., Lewis, J., Raff, M., Roberts, K., and Watson, J. D., 1989, *Molecular Biology of the Cell*, Garland Publishing, Inc., New York.

Alifano, P., Bruni, C. B., and Carlomagno, M. S., 1994, Control of mRNA processing and decay in prokaryotes, *Genetica* **94**:157.

Ansari, A., Kuznetsov, S. V., and Shen, Y., 2001, Configurational diffusion down a folding funnel describes the dynamics of DNA hairpins, *Proc. Natl. Acad. Sci. USA* **98**:7771.

Ansari, A., Shen, Y., and Kuznetsov, S. V., 2002, Misfolded loops decrease the effective rate of DNA hairpin formation, *Phys. Rev. Lett.* **88**:069801.

Armitage, B., Ly, D., Koch, T., Frydenlund, H., Orum, H., and Schuster, G. B., 1998, Hairpin-forming peptide nucleic acid oligomers, *Biochemistry* **37**:9417.

Baldwin, R. L., 1996, Why is protein folding so fast?, *Proc. Natl. Acad. Sci. USA* **93**:2627.

Bianco, P. R., Brewer, L. R., Corzett, M., Balhorn, R., Yeh, Y., Kowalczykowski, S. C., and Baskin, R. J., 2001, Processive translocation and DNA unwinding by individual RecBCD enzyme molecules, *Nature* **409**:374.

Bieri, O., Wirz, J., Hellrung, B., Schutkowski, M., Drewello, M., and Kiefhaber, T., 1999, The speed limit for protein folding measured by triplet-triplet energy transfer, *Proc. Natl. Acad. Sci. USA* **96**:9597.

Bockelmann, U., Thomen, P., Essevaz-Roulet, B., Viasnoff, V., and Heslot, F., 2002, Unzipping DNA with optical tweezers: high sequence sensitivity and force flips, *Biophys. J.* **82**:1537.

Bonnet, G., Krichevsky, O., and Libchaber, A., 1998, Kinetics of conformational fluctuations in DNA hairpin-loops, *Proc. Natl. Acad. Sci. USA* **95**:8602.

Bonnet, G., Tyagi, S., Libchaber, A., and Kramer, F. R., 1999, Thermodynamic basis of the enhanced specificity of structured DNA probes, *Proc. Natl. Acad. Sci. USA* **96**:6171.

Bryngelson, J. D., Onuchic, J. N., Socci, N. D., and Wolynes, P. G., 1995, Funnels, pathways, and the energy landscape of protein folding: a synthesis, *Proteins* **21**:167.

Bryngelson, J. D., and Wolynes, P. G., 1989, Intermediates and barrier crossing in a random energy model (with applications to protein folding), *J. Phys. Chem.* **93**:6902.

Bustamante, C., Marko, J. F., Siggia, E. D., and Smith, S., 1994, Entropic elasticity of lambda-phage DNA, *Science* **265**:1599.

Bustamante, C., Smith, S. B., Liphardt, J., and Smith, D., 2000, Single-molecule studies of DNA mechanics, *Curr. Opin. Struct. Biol.* **10**:279.

Cantor, C. R., and Schimmel, P. R., 1980, *The behavior of biological macromolecules*, W. H. Freeman and Company, New York.

Caskey, C. T., Pizzuti, A., Fu, Y. H., Fenwick, R. G. J., and Nelson, D. L., 1992, Triplet repeat mutations in human disease, *Science* **256**:784.

Cech, T. R., Zaug, A. J., and Grabowski, P. J., 1981, In vitro splicing of the ribosomal RNA precursor of Tetrahymena: involvement of a guanosine nucleotide in the excision of the intervening sequence, *Cell.* **27**:487.

Chalikian, T. V., Volker, J., Plum, G. E., and Breslauer, K. J., 1999, A more unified picture for the thermodynamics of nucleic acid duplex melting: a characterization by calorimetric and volumetric techniques, *Proc. Natl. Acad. Sci. USA* **96**:7853.

Chen, S. J., and Dill, K. A., 2000, RNA folding energy landscapes, *Proc. Natl. Acad. Sci. USA* **97**:646.

Cheng, J. C., Moore, T. B., and Sakamoto, K. M., 2003, RNA interference and human disease., *Mol. Genet. Metab.* **80**:121.

Cheong, C., Varani, G., and Tinoco, I. J., 1990, Solution structure of an unusually stable RNA hairpin, 5'GGAC(UUCG)GUCC, *Nature* **346**:680.

Chu, Y. G., and Tinoco, I., 1983, Temperature-jump kinetics of the dC-G-T-G-A-A-T-T-C-G-C-G double helix containing a G.T base-pair and the dC-G-C-A-G-A-A-T-T-C-G-C-G double helix containing an extra adenine, *Biopolymers* **22**:1235.

Cocco, S., Marko, J. F., and Monasson, R., 2003a, Slow nucleic acid unzipping kinetics from sequence-defined barriers, *Eur. Phys. J. E.* **10**:153.

Cocco, S., Marko, J. F., Monasson, R., Sarkar, A., and Yan, J., 2003b, Force-extension behavior of folding polymers, *Eur. Phys. J. E.* **10**:249.

Cohen, R. J., and Crothers, D. M., 1971, Rate of unwinding small DNA, *J. Mol. Biol.* **61**:525.

Coutts, S. M., 1971, Thermodynamics and kinetics of G-C base pairing in the isolated extra arm of serine-specific transfer RNA from yeast, *Biochim. Biophys. Acta* **232**:94.

Craig, M. E., Crothers, D. M., and Doty, P., 1971, Relaxation kinetics of dimer formation by self complementary oligonucleotides, *J. Mol. Biol.* **62**:383.

Crews, S., Ojala, D., Posakony, J., Nishiguchi, J., and Attardi, G., 1979, Nucleotide sequence of a region of human mitochondrial DNA containing the precisely identified origin of replication, *Nature* **277**:192.

Dai, X., Greizerstein, M. B., Nadas-Chinni, K., and Rothman-Denes, L. B., 1997, Supercoil-induced extrusion of a regulatory DNA hairpin, *Proc. Natl. Acad. Sci. USA* **94**:2174.

Dai, X., Kloster, M., and Rothman-Denes, L. B., 1998, Sequence-dependent extrusion of a small DNA hairpin at the N4 virion RNA polymerase promoters, *J. Mol. Biol.* **283**:43.

DeGennes, P. G., 1979, *Scaling Concepts in Polymer Physics*, Cornell University Press, Ithaca, N.Y.

Deniz, A. A., Dahan, M., Grunwell, J. R., Ha, T., Faulhaber, A. E., Chemla, D. S., Weiss, S., and Schultz, P. G., 1999, Single-pair fluorescence resonance energy transfer on freely diffusing molecules: observation of Forster distance dependence and subpopulations, *Proc. Natl. Acad. Sci. USA* **96**:3670.

Dill, K. A., 1990, Dominant forces in protein folding, *Biochemistry* **29**:7133.

Dill, K. A., and Bromberg, S., 2003, *Molecular Driving Forces: Statistical Thermodynamics in Chemistry and Biology*, Garland Science Publishing, New York.

Dill, K. A., and Chan, H. S., 1997, From Levinthal to pathways to funnels, *Nat. Struct. Biol.* **4**:10.

Doi, M., 1975, Diffusion-controlled reaction of polymers, *Chem. Phys.* **9**:455.

Doudna, J. A., and Doherty, E. A., 1997, Emerging themes in RNA folding, *Fold Des.* **2**:R65.

Draper, D. E., 1999, Themes in RNA-protein recognition, *J. Mol. Biol.* **293**:255.

Dyer, R. B., Gai, F., Woodruff, W. H., Gilmanshin, R., and Callender, R. H., 1998, Infrared studies of fast events in protein folding, *Acc. Chem. Res.* **31**:709.

Dykxhoorn, D. M., Novina, C. D., and Sharp, P. A., 2003, Killing the messenger: short RNAs that silence gene expression, *Nature Rev. Mol. Cell Biol.* **4**:457.

Eaton, W. A., Munoz, V., Thompson, P. A., Henry, E. R., and Hofrichter, J., 1998, Kinetics and dynamics of loops, alpha-helices, β-hairpins, and fast-folding proteins, *Acc. Chem. Res.* **31**:745.

Eigen, M., and de Maeyer, L., 1963, *Relaxation Methods*, Interscience, New York, 895.

Eisenberg, H., and Felsenfeld, G., 1967, Studies of the temperature-dependent conformation and phase separation of polyriboadenylic acid solutions at neutral pH, *J. Mol. Biol.* **30**:17.

Essevaz-Roulet, B., Bockelmann, U., and Heslot, F., 1997, Mechanical separation of the complementary strands of DNA, *Proc. Natl. Acad. Sci. USA* **94**:11935.

Fire, A., Xu, S., Montgomery, M. K., Kostas, S. A., Driver, S. E., and Mello, C. C., 1998, Potent and specific genetic interference by double-stranded RNA in Caenorhabditis elegans, *Nature* **391**:806.

Flamm, C., Fontana, W., Hofacker, I. L., and Schuster, P., 2000, RNA folding at elementary step resolution, *RNA* **6**:325.

Friedman, B., and O'Shaughnessy, B., 1989, Theory of polymer cyclization, *Phys. Rev. A.* **40**:5950.

Gacy, A. M., Goellner, G., Juranic, N., Macura, S., and McMurray, C. T., 1995, Trinucleotide repeats that expand in human disease form hairpin structures in vitro, *Cell* **81**:533.

Gerland, U., Bundschuh, R., and Hwa, T., 2001, Force-induced denaturation of RNA, *Biophys. J.* **81**:1324.

Glucksmann-Kuis, M. A., Dai, X., Markiewicz, P., and Rothman-Denes, L. B., 1996, E. coli SSB activates N4 virion RNA polymerase promoters by stabilizing a DNA hairpin required for promoter recognition, *Cell* **84**:147.

Goddard, N. L., Bonnet, G., Krichevsky, O., and Libchaber, A., 2000, Sequence dependent rigidity of single-stranded DNA, *Phys. Rev. Lett.* **85**:2400.

Goddard, N. L., Bonnet, G., Krichevsky, O., and Libchaber, A., 2002, Goddard et al. reply, *Phys. Rev. Lett.* **88**:069802.

Gralla, J., and Crothers, D. M., 1973, Free energy of imperfect nucleic acid helices. II. Small hairpin loops, *J. Mol. Biol.* **73**:497.

Gregorian, R. S., Jr., and Crothers, D. M., 1995, Determinants of RNA hairpin loop-loop complex stability, *J. Mol. Biol.* **248**:968.

Gruebele, M., Sabelko, J., Ballew, R., and Ervin, J., 1998, Laser temperature jump induced protein refolding, *Acc. Chem. Res.* **31**:699.

Grunwell, J. R., Glass, J. L., Lacoste, T. D., Deniz, A. A., Chemla, D. S., and Schultz, P. G., 2001, Monitoring the conformational fluctuations of DNA hairpins using single-pair fluorescence resonance energy transfer, *J. Am. Chem. Soc.* **123**:4295.

Guo, Z., and Thirumalai, D., 1995, Kinetics of protein folding: Nucleation mechanism, time scales, and pathways, *Biopolymers* **36**:83.

Hagen, S. J., Carswell, C. W., and Sjolander, E. M., 2001, Rate of intrachain contact formation in an unfolded protein: temperature and denaturant effects, *J. Mol. Biol.* **305**:1161.

Hagen, S. J., Hofrichter, J., Szabo, A., and Eaton, W. A., 1996, Diffusion-limited contact formation in unfolded cytochrome c: estimating the maximum rate of protein folding, *Proc. Natl. Acad. Sci. USA* **93**:11615.

Hagerman, P. J., 1988, Flexibility of DNA, *Annu. Rev. Biophys. Biophys. Chem.* **17**:265.

Hamilton, A. J., and Baulcombe, D. C., 1999, A species of small antisense RNA in posttranscriptional gene silencing in plants, *Science* **286**:950.

Heus, H. A., and Pardi, A., 1991, Structural features that give rise to the unusual stability of RNA hairpins containing GNRA loops, *Science* **253**:191.

Hoffman, G. W., 1971, A Nanosecond Temperature-Jump Apparatus, *Rev. Sci. Instruments* **42**:1643.

Hofrichter, J., 2001, Laser Temperature-Jump Methods for Studying Folding Dynamics, in: *Methods in Molecular Biology*, K. P. Murphy, ed., Humana Press, Totowa, New Jersey.

Hudgins, R. R., Huang, F., Gramlich, G., and Nau, W. M., 2002, A fluorescence-based method for direct measurement of submicrosecond intramolecular contact formation in biopolymers: an exploratory study with polypeptides, *J. Am. Chem. Soc.* **124**:556.

Inners, L. D., and Felsenfeld, G., 1970, Conformation of polyribouridylic acid in solution, *J. Mol. Biol.* **50**:373.

Isambert, H., and Siggia, E. D., 2000, Modeling RNA folding paths with pseudoknots: application to hepatitis delta virus ribozyme, *Proc. Natl. Acad. Sci. USA* **97**:6515.

Jacob, M., Schindler, T., Balbach, J., and Schmid, F. X., 1997, Diffusion control in an elementary protein folding reaction, *Proc. Natl. Acad. Sci. USA* **94**:5622.

Jaeger, J. A., SantaLucia, J., Jr., and Tinoco, I., Jr., 1993, Determination of RNA structure and thermodynamics, *Annu. Rev. Biochem.* **62**:255.

Jaeger, J. A., Turner, D. H., and Zuker, M., 1990, Predicting optimal and suboptimal secondary structure for RNA, *Methods Enzymol.* **183**:281.

Jager, M., Nguyen, H., Crane, J. C., Kelly, J. W., and Gruebele, M., 2001, The folding mechanism of a β–sheet: The WW domain, *J. Mol. Biol.* **311**:373.

Klausner, R. D., Rouault, T. A., and Harford, J. B., 1993, Regulating the fate of mRNA: the control of cellular iron metabolism, *Cell.* **72**:19.

Kuznetsov, S. V., Kozlov, A. G., Lohman, T. M., and Ansari, A., 2004, Kinetics of wrapping/unwrapping of single-stranded DNA around the *Escherichia coli* SSB tetramer, *Biophys. J.* **86**:588a.

Kuznetsov, S. V., Shen, Y., Benight, A. S., and Ansari, A., 2001, A semiflexible polymer model applied to loop formation in DNA hairpins, *Biophys. J.* **81**:2864.

Landau, L. D., and Lifshitz, E. M., 1980, *Statistical Physics*, Pergamon Press, New York.

Lapidus, L. J., Eaton, W. A., and Hofrichter, J., 2000, Measuring the rate of intramolecular contact formation in polypeptides, *Proc. Natl. Acad. Sci. USA* **97**:7220.

Levinthal, C., 1969, How to Fold Graciously?, in: *Mossbauer Spectroscopy in Biological Systems: Proceedings of a meeting held at Allerton House, Monticello, Illinois.*, P. DeBrunner, J. C. M. Tsibris and E. Munck, eds., Univ. of Illinois Bulletin, Urbana, IL, pp. 22-24.

Lilley, D. M., 1981, Hairpin-loop formation by inverted repeats in supercoiled DNA is a local and transmissible property, *Nucleic Acids Res.* **9**:1271.

Lilley, D. M., 2003, The origins of RNA catalysis in ribozymes, *Trends Biochem. Sci.* **28**:495.

Liphardt, J., Onoa, B., Smith, S. B., Tinoco, I. J., and Bustamante, C., 2001, Reversible unfolding of single RNA molecules by mechanical force, *Science* **292**:733.

Lohman, T. M., and Bjornson, K. P., 1996, Mechanisms of helicase-catalyzed DNA unwinding, *Annu. Rev. Biochem.* **65**:169.

Maier, B., Bensimon, D., and Croquette, V., 2000, Replication by a single DNA polymerase of a stretched single-stranded DNA, *Proc. Natl. Acad. Sci. USA* **97**:12002.

Marino, J. P., Gregorian, R. S., Csankovszki, G., and Crothers, D. M., 1995, Bent helix formation between RNA hairpins with complementary loops, *Science* **268**:1448.

Marko, J. F., and Siggia, E. D., 1994, Fluctuations and supercoiling of DNA, *Science* **265**:506.

Mathews, D. H., Sabina, J., Zuker, M., and Turner, D. H., 1999, Expanded sequence dependence of thermodynamic parameters improves prediction of RNA secondary structure, *J. Mol. Biol.* **288**:911.

Miller, J. L., and Kollman, P. A., 1997a, Observation of an A-DNA to B-DNA transition in a nonhelical nucleic acid hairpin molecule using molecular dynamics, *Biophys. J.* **73**:2702.

Miller, J. L., and Kollman, P. A., 1997b, Theoretical studies of an exceptionally stable RNA tetraloop: observation of convergence from an incorrect NMR structure to the correct one using unrestrained molecular dynamics, *J. Mol. Biol.* **270**:436.

Miller, W. G., Brant, D. A., and Flory, P. J., 1967, Random coil configurations of polypeptide copolymers, *J. Mol. Biol.* **23**:67.

Mills, J. B., Vacano, E., and Hagerman, P. J., 1999, Flexibility of single-stranded DNA: use of gapped duplex helices to determine the persistence lengths of poly(dT) and poly(dA), *J. Mol. Biol.* **285**:245.

Montanari, A., and Mezard, M., 2001, Hairpin formation and elongation of biomolecules, *Phys. Rev. Lett.* **86**:2178.

Munoz, V., and Eaton, W. A., 1999, A simple model for calculating the kinetics of protein folding from three-dimensional structures, *Proc. Natl. Acad. Sci. USA* **96**:11311.

Munoz, V., Thompson, P. A., Hofrichter, J., and Eaton, W. A., 1997, Folding dynamics and mechanism of β-hairpin formation, *Nature* **390**:196.

Nielsen, P. E., 1999, Peptide nucleic acids as therapeutic agents, *Current opinion in structural biology*. **9**:353.

Onuchic, J. N., Luthey-Schulten, Z., and Wolynes, P. G., 1997, Theory of protein folding: the energy landscape perspective, *Annu. Rev. Phys. Chem.* **48**:545.

Pan, J., Thirumalai, D., and Woodson, S. A., 1997, Folding of RNA involves parallel pathways, *J. Mol. Biol.* **273**:7.

Pan, T., and Sosnick, T. R., 1997, Intermediates and kinetic traps in the folding of a large ribozyme revealed by circular dichroism and UV absorbance spectroscopies and catalytic activity, *Nat. Struct. Biol.* **4**:931.

Plaxco, K. W., and Baker, D., 1998, Limited internal friction in the rate-limiting step of a two-state protein folding reaction, *Proc. Natl. Acad. Sci. USA* **95**:13591.

Podtelezhnikov, A., and Vologodskii, A., 1997, Simulations of polymer cyclization by brownian dynamics, *Macromolecules* **30**:6668.

Poland, D., and Scheraga, H. A., 1970, *Theory of helix-coil transitions in biopolymers; statistical mechanical theory of order-disorder transitions in biological macromolecules*, Academic Press, New York,, xvii.

Porschke, D., 1974a, A direct measure of the unzippering rate of a nucleic acid double helix, *Biophys. Chem.* **2**:97.

Porschke, D., 1974b, Thermodynamics and Kinetics Parameters of an Oligonucleotide Hairpin Helix, *Biophys. Chem.* **1**:381.

Porschke, D., 1977, Elementary steps of base recognition and helix-coil transitions in nucleic acids, *Mol. Biol. Biochem. Biophys.* **24**:191.

Porschke, D., and Eigen, M., 1971, Co-operative non-enzymic base recognition. 3. Kinetics of the helix- coil transition of the oligoribouridylic--oligoriboadenylic acid system and of oligoriboadenylic acid alone at acidic pH, *J. Mol. Biol.* **62**:361.

Porschke, D., Uhlenbeck, O. C., and Martin, F. H., 1973, Thermodynamics and kinetics of the helix-coil transition of oligomers containing GC base pairs., *Biopolymers* **12**:1313.

Rief, M., Clausen-Schaumann, H., and Gaub, H. E., 1999, Sequence-dependent mechanics of single DNA molecules, *Nat. Struct. Biol.* **6**:346.

Rivetti, C., Walker, C., and Bustamante, C., 1998, Polymer chain statistics and conformational analysis of DNA molecules with bends or sections of different flexibility, *J. Mol. Biol.* **280**:41.

Romer, R., Schomburg, U., Krauss, G., and Maass, G., 1984, Escherichia coli single-stranded DNA binding protein is mobile on DNA: 1H NMR study of its interaction with oligo- and polynucleotides, *Biochemistry* **23**:6132.

Roth, D. B., Menetski, J. P., Nakajima, P. B., Bosma, M. J., and Gellert, M., 1992, V(D)J recombination: broken DNA molecules with covalently sealed (hairpin) coding ends in scid mouse thymocytes, *Cell* **70**:983.

Rouzina, I., and Bloomfield, V. A., 1999, Heat capacity effects on the melting of DNA. 1. General aspects, *Biophys. J.* **77**:3242.

Rupert, P. B., Massey, A. P., Sigurdsson, S. T., and Ferre-D'Amare, A. R., 2002, Transition state stabilization by a catalytic RNA, *Science* **298**:1421.

Sali, A., Shakhnovich, E., and Karplus, M., 1994a, How does a protein fold?, *Nature* **369**:248.

Sali, A., Shakhnovich, E., and Karplus, M., 1994b, Kinetics of protein folding. A lattice model study of the requirements for folding to the native state, *J. Mol. Biol.* **235**:1614.

Sarzynska, J., Kulinski, T., and Nilsson, L., 2000, Conformational dynamics of a 5S rRNA hairpin domain containing loop D and a single nucleotide bulge, *Biophys. J.* **79**:1213.

Schindler, T., and Schmid, F. X., 1996, Thermodynamic properties of an extremely rapid protein folding reaction, *Biochemistry* **35**:16833.

Sclavi, B., Sullivan, M., Chance, M. R., Brenowitz, M., and Woodson, S. A., 1998, RNA folding at millisecond intervals by synchrotron hydroxyl radical footprinting, *Science* **279**:1940.

Serra, M. J., and Turner, D. H., 1995, Predicting thermodynamic properties of RNA, *Methods Enzymol.* **259**:242.

Shen, Y., Kuznetsov, S. V., and Ansari, A., 2001, Loop dependence of the dynamics of DNA hairpins, *J. Phys. Chem. B.* **105**:12202.

Shirts, M. R., and Pande, V. S., 2001, Mathematical analysis of coupled parallel simulations, *Phys. Rev. Lett.* **86**:4983.

Smith, S. B., Cui, Y., and Bustamante, C., 1996, Overstretching B-DNA: the elastic response of individual double- stranded and single-stranded DNA molecules, *Science* **271**:795.

Smolke, C. D., Carrier, T. A., and Keasling, J. D., 2000, Coordinated, differential expression of two genes through directed mRNA cleavage and stabilization by secondary structures, *Appl. Environ. Microbiol.* **66**:5399.

Socci, N. D., Onuchic, J. N., and Wolynes, P. G., 1996, Diffusive dynamics of the reaction coordinate for protein folding funnels, *J. Chem. Phys.* **104**:5860.

Sorin, E. J., Engelhardt, M. A., Herschlag, D., and Pande, V. S., 2002, RNA simulations: probing hairpin unfolding and the dynamics of a GNRA tetraloop, *J. Mol. Biol.* **317**:493.

Sorin, E. J., Rhee, Y. M., Nakatani, B. J., and Pande, V. S., 2003, Insights into nucleic acid conformational dynamics from massively parallel stochastic simulations, *Biophys. J.* **85**:790.

Spatz, H., and Baldwin, R. L., 1965, Study of the folding of the dAT copolymer by kinetics measurements of melting, *J. Mol. Biol.* **11**:213.

Srinivasan, J., Miller, J., Kollman, P. A., and Case, D. A., 1998, Continuum solvent studies of the stability of RNA hairpin loops and helices, *J. Biomol. Struct. Dyn.* **16**:671.

Stannard, B. S., and Felsenfeld, G., 1975, The conformation of polyriboadenylic acid at low temperature and neutral pH. A single-stranded rodlike structure, *Biopolymers* **14**:299.

Steinfeld, J. I., Francisco, J. S., and Haas, W. L., 1999, *Chemical Kinetics and Dynamics*, Prentice Hall, New Jersey.

Szabo, A., Schulten, K., and Schulten, Z., 1980, First passage time approach to diffusion controlled reactions, *J. Chem. Phys.* **72**:4350.

Tan, Y. J., Oliveberg, M., and Fersht, A. R., 1996, Titration properties and thermodynamics of the transition state for folding: comparison of two-state and multi-state folding pathways, *J. Mol. Biol.* **264**:377.

Thirumalai, D., 1998, Native secondary structure formation in RNA may be a slave to tertiary folding, *Proc. Natl. Acad. Sci. USA* **95**:11506.

Thirumalai, D., Lee, N., Woodson, S. A., and Klimov, D., 2001, Early events in RNA folding, *Annu. Rev. Phys. Chem.* **52**:751.

Thirumalai, D., and Woodson, S. A., 1996, Kinetics of folding of proteins and RNA, *Acc. Chem. Res.* **29**:433.

Thomen, P., Bockelmann, U., and Heslot, F., 2002, Rotational drag on DNA: a single molecule experiment, *Phys. Rev. Lett.* **88**:248102.

Tinland, B., Pluen, A., Sturm, J., and Weill, G., 1997, Persistence length of single-stranded DNA, *Macromolecules* **30**:5763.

Turner, D. H., 2000, Conformational Changes, in: *Nucleic Acids: Structures, Properties, and Functions*, V. A. Bloomfield, D. M. Crothers and I. J. Tinoco, ed., University Science Books, Sausalito, California.

Turner, D. H., and Sugimoto, N., 1988, RNA structure prediction, *Annu. Rev. Biophys. Biophys. Chem.* **17**:167.

Uhlenbeck, O. C., 1990, Tetraloops and RNA folding, *Nature* **346**:613.

Uhlenbeck, O. C., Borer, P. N., Dengler, B., and Tinoco, I., Jr., 1973, Stability of RNA hairpin loops: A_6-C_m-U_6, *J. Mol. Biol.* **73**:483.

Vallone, P. M., and Benight, A. S., 1999, Melting studies of short DNA hairpins containing the universal base 5- nitroindole, *Nucleic Acids Res.* **27**:3589.

Vallone, P. M., Paner, T. M., Hilario, J., Lane, M. J., Faldasz, B. D., and Benight, A. S., 1999, Melting studies of short DNA hairpins: influence of loop sequence and adjoining base pair identity on hairpin thermodynamic stability, *Biopolymers* **50**:425.

Varani, G., 1995, Exceptionally stable nucleic acid hairpins, *Annu. Rev. Biophys. Biomol. Struct.* **24**:379.

Vesnaver, G., and Breslauer, K. J., 1991, The contribution of DNA single-stranded order to the thermodynamics of duplex formation, *Proc. Natl. Acad. Sci. USA* **88**:3569.

Wallace, M. I., Ying, L., Balasubramanian, S., and Klenerman, D., 2000, FRET fluctuation spectroscopy: exploring the conformational dynamics of a DNA hairpin loop, *J. Phys. Chem. B.* **104**:11551.

Wallace, M. I., Ying, L., Balasubramanian, S., and Klenerman, D., 2001, Non-Arrhenius kinetics for the loop closure of a DNA hairpin, *Proc. Natl. Acad. Sci. USA* **98**:5584.

Wang, J. C., and Davidson, N., 1966a, On the probability of ring closure of lambda DNA, *J. Mol. Biol.* **19**:469.

Wang, J. C., and Davidson, N., 1966b, Thermodynamic and kinetic studies on the interconversion between the linear and circular forms of phage lambda DNA, *J. Mol. Biol.* **15**:111.

Wang, J. C., and Davidson, N., 1968, Cyclization of phage DNAs, *Cold. Spring Harb. Symp. Quant. Biol.* **33**:409.

Wartell, R. M., and Benight, A. S., 1985, Thermal denaturation of DNA molecules: a comparison of theory with experiment, *Phys. Rep.* **126**:67.

Wetmur, J. G., and Davidson, N., 1968, Kinetics of renaturation of DNA, *J. Mol. Biol.* **31**:349.

Wilemski, G., and Fixman, M., 1974, Diffusion-controlled intrachain reactions of polymers. I. Theory, *J. Chem. Phys.* **60**:866.

Williams, A. P., Longfellow, C. E., Freier, S. M., Kierzek, R., and Turner, D. H., 1989, Laser temperature-jump, spectroscopic, and thermodynamic study of salt effects on duplex formation by dGCATGC, *Biochemistry* **28**:4283.

Williams, D. J., and Hall, K. B., 1999, Unrestrained stochastic dynamics simulations of the UUCG tetraloop using an implicit solvation model, *Biophys. J.* **76**:3192.

Williams, M. C., Wenner, J. R., Rouzina, I., and Bloomfield, V. A., 2001, Entropy and heat capacity of DNA melting from temperature dependence of single molecule stretching, *Biophys. J.* **80**:1932.

Wilson, K. S., and von Hippel, P. H., 1995, Transcription termination at intrinsic terminators: the role of the RNA hairpin, *Proc. Natl. Acad. Sci. USA* **92**:8793.

Winnik, M. A., 1986, *Cyclic Polymers*, Elsevier, New York.

Wolynes, P. G., Onuchic, J. N., and Thirumalai, D., 1995, Navigating the folding routes, *Science* **267**:1619.

Wu, M., and Tinoco, I., Jr., 1998, RNA folding causes secondary structure rearrangement, *Proc. Natl. Acad. Sci. USA* **95**:11555.

Wuite, G. J., Smith, S. B., Young, M., Keller, D., and Bustamante, C., 2000, Single-molecule studies of the effect of template tension on T7 DNA polymerase activity, *Nature* **404**:103.

Xodo, L. E., Manzini, G., Quadrifoglio, F., van der Marel, G. A., and van Boom, J. H., 1988, The duplex-hairpin conformational transition of d(CGCGCGATCGCGCG) and d(CGCGCGTACGCGCG): a thermodynamic and kinetic study, *J. Biomol. Struct. Dyn.* **6**:139.

Young, M. A., Ravishanker, G., and Beveridge, D. L., 1997, A 5-nanosecond molecular dynamics trajectory for B-DNA: analysis of structure, motions, and solvation, *Biophys. J.* **73**:2313.

Zacharias, M., 2001, Conformational analysis of DNA-trinucleotide-hairpin-loop structures using a continuum solvent model, *Biophys. J.* **80**:2350.

Zagrovic, B., Sorin, E. J., and Pande, V., 2001, β-hairpin folding simulations in atomistic detail using an implicit solvent model, *J. Mol. Biol.* **313**:151.

Zarrinkar, P. P., Wang, J., and Williamson, J. R., 1996, Slow folding kinetics of RNase P RNA, *RNA* **2**:564.

Zarrinkar, P. P., and Williamson, J. R., 1994, Kinetic intermediates in RNA folding, *Science* **265**:918.

Zhang, W., and Chen, S. J., 2002, RNA Hairpin Folding Kinetics, *Proc. Natl. Acad. Sci. USA* **99**:1931.

Zhang, Y., Zhou, H., and Ou-Yang, Z. C., 2001, Stretching single-stranded DNA: interplay of electrostatic, base-pairing, and base-pair stacking interactions, *Biophys. J.* **81**:1133.

Zhuang, X., Bartley, L. E., Babcock, H. P., Russell, R., Ha, T., Herschlag, D., and Chu, S., 2000, A single-molecule study of RNA catalysis and folding, *Science* **288**:2048.

Zichi, D. A., 1995, Molecular dynamics of RNA with the OPLS force field. Aqueous simulations of a hairpin containing a tetranucleotide loop, *J. Am. Chem. Soc.* **117**:2957.
Zuker, M., 1989, On finding all suboptimal foldings of an RNA molecule, *Science* **244**:48.
Zwanzig, R., 1988, Diffusion in a rough potential, *Proc. Natl. Acad. Sci. USA* **85**:2029.
Zwanzig, R., Szabo, A., and Bagchi, B., 1992, Levinthal's paradox, *Proc. Natl. Acad. Sci. USA* **89**:20.

BIOINSPIRED APPROACHES TO BUILDING NANOSCALE DEVICES

Sawitri Mardyani, Wen Jiang, Jonathan Lai, Jane Zhang, Warren C. W. Chan*

1. INTRODUCTION

In 1959, Richard Feynman gave a speech called "There's plenty of room at the bottom," where he challenged scientists to construct atomic-scale devices. Since 1959, the dream of building nanoscale devices (~ 1 x 10 $^{-9}$ m) is slowly becoming a reality. The expected impact of nanodevices is immense. Nanodevices are comparable in size to most biological systems (e.g., viruses, organelles, proteins, DNA) and therefore, it has been suggested that nanodevices can be engineered to diagnose and treat malfunctions in cells. This can lead to a new generation of treatment strategies for cancer, AIDS, or Alzheimer. Future advances in the computer chip industry may also depend on nanotechnology – the ability to design atomic-scale chips may provide hard drives with greater memory capabilities or faster electronics. Assembled nanosystems are also expected to impact the aerospace program – where lightweight, high tear-resistant fabric (made with nanostructures) with embedded nanosensors may be designed for astronauts. To meet the challenges and rewards of nanotechnology, research institutions around the world have built infrastructures to study, manipulate, and design nanometer-sized materials (generally, < 100 nm) for building atomic-scale devices.

In the last couple of centuries, we have perfected the art of building macroscale devices such as clocks, calculators, computer chips, and cars. The building process generally involves the manufacturing of precursor components and the assembling of these components into a functional unit. The entire process can be automated using man-made machines. However, there is extreme difficulty in building similar devices with nanometer-dimensions. This is mainly due to the inability to assemble atoms or nanometer-sized components in a coordinated fashion. Top-down approaches, generally associated with photolithography and etching techniques, have been instrumental in advances in microscale technologies. However, a top-down approach has found limited

* Sawitri Mardyani, Wen Jian, Jonathan Lai, Jane Zhang, and Warren C. W. Chan, Institute of Biomaterials and Biomedical Engineering, University of Toronto, Toronto, Ontario, Canada M5S 3G9. Warren C. W. Chan is also affiliated with Department of Materials Science and Engineering, University of Toronto.

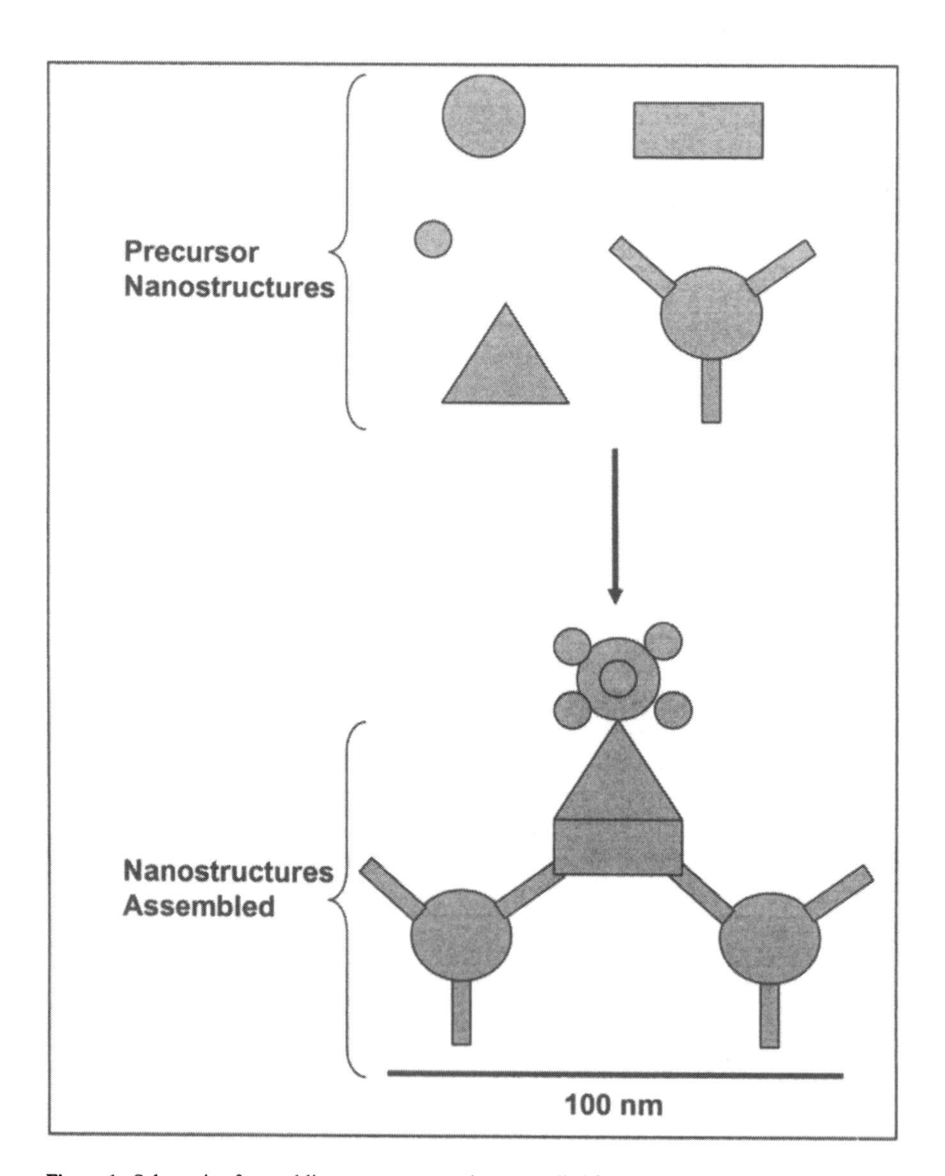

Figure 1. Schematic of assembling nanostructures in a controlled fashion.

use in building nanoscale devices mainly due to high costs of instruments and diffraction limit. A bottom-up strategy, where atoms are precisely assembled in the molecular scale, is one of the most promising strategies for building atomic-scale devices, as illustrated in Figure 1. Material scientists and chemists have spent the last thirty years perfecting the synthesis of nanoscale objects, that may one day become precursors for building nanoscale devices.[1-6] Semiconductor, metallic, non-metallic, and alloyed nanoparticles have been successfully synthesized and characterized. The precision of synthesis is so great that researchers can selectively and reproducibly grow protrusions on the surface of spherical quantum dots[7] or modify the dimensions of metallic nanoparticles using solution-based approaches.[8, 9] Simple monofunctional hybrid organic/inorganic nanostructures have demonstrated a plethora of applications in biomedical detection. Yet, the current challenge of nanotechnologists is to mix-and-match different types of nanostructures in a coordinated fashion to produce a functional device the size of a virus (< 100 nm).

Integrating organic molecules with inorganic nanostructures have produced exciting results in the field of nanodevice building. Organic molecules have been utilized as surface coatings to prevent unwanted nanoparticle aggregation, as molecules to direct nanoparticle assembly, and as homing devices to target nanostructures to specific biological sites. Furthermore, organic molecules have embarked greater functional capabilities into inorganic nanostructures. For example, Mirkin and coworkers used oligonucleotides to assemble and de-assemble metallic nanostructures[10, 11]. Montemagno and coworkers programmed rotational motion in nickel nanoparticles[12, 13] while Ruoslahti and coworkers directed quantum dots to tumour sites using homing peptides *in vivo.*[14] At the current state of research in this field, there is a wide-array of precursor nanostructures and organic molecules available to build nanodevices. However, the key challenge is to develop novel approaches to assemble them into a functional unit. In this chapter, we will describe some of the precursors available for device building and describe some of the recent strategies that utilize biological systems or biomimetic systems to assemble nanostructures.

2. NANOSTRUCTURES AS CORE COMPONENTS FOR BUILDING DEVICES

In the last thirty years, great efforts in the synthesis of nanoscale materials have provided researchers with a large set of building blocks for constructing nanoscale devices. Materials in this size-regime possess unique optical, electronic, and magnetic properties that can be tuned by altering the particle's size, shape, or composition. Metallic nanostructures have the ability to produce heat, quench luminescence, scatter light, and create surface plasmon upon optical excitation. Semiconductor nanostructures have the ability to emit light and are conductive upon optical or electrical excitation. Rod-shaped semiconductor nanostructures can polarize light while spherical shaped nanostructures cannot. Carbon nanotubes have unique electrical and mechanical properties, and are lightweight. Fullerenes are hollow in the core while dendrimers are highly porous. These are examples of some of the properties of nanostructures that may find use in nanoscale device building.

The ability to assemble nanostructures requires precise control of the particle's surface chemistry, where molecules can be coated onto the surface to direct the assembly process. Strategies have been developed to readily permit the modification of a nanoparticle's surface chemistry.[15-19] The easiest types of nanoparticles to coat with recognition biomolecules are metallic nanostructures such as gold or silver colloids. In preparing for coating, the surface of metallic nanoparticles is generally stabilized with a weak ligand that can be easily desorbed from the surface. For example, proteins such as bovine serum albumin and transferrin can be adsorbed onto the surface of citrate stabilized gold nanoparticles through ionic and hydrophobic interactions as well as dative binding. For other types of nanoparticles where the surface coating with biomolecules may be more difficult, extra processing steps are needed to create a surface with reactive functional groups (-COOH, -SH, or $-NH_2$). For example, semiconductor quantum dots do not readily adsorb proteins. In order to coat the surface of quantum dots with proteins, a layer of amphipolic polymer can be used to render the surface of the quantum dots with -COOH functional groups.[20] Primary $-NH_2$ group from biomolecules can then be cross-linked to the surface of the amphipolic polymer through a carbodiimide-assisted reaction, forming an amide bond linkage between the quantum dot and biomolecule. Specialized techniques have been developed for functionalizing carbon nanotubes, magnetic nanoparticles, and fullerenes for cross-linking to biomolecules.

3. BIOLOGY AS MODEL SYSTEM FOR BUILDING NANOSCALE DEVICES

A key challenge in building nanoscale devices is the ability to assemble components in a controlled fashion. How can one build a nanometer-sized RC-circuit? How can one build a detection and drug storage and delivery system the size of a standard virus (~50 to 150 nm)? Researchers have recently looked toward biology as a guide to assemble nanostructures into functional devices. One interesting fact about bioassembly in living cells and tissues is that only a small amount of subunits are required to produce a rich and diverse group of functional systems that regulate and maintain the viability of cells, tissues, organs, and the organism. The information in DNA, the blueprint of a cell, is encoded using only four distinct bases. Proteins, the functional units of a cell, are only composed of 20-amino acids. Yet, a cell has developed the capability to assemble 4-bases into a complete genetic code and 20-different amino acids into thousands of functional proteins.

The molecular inner-workings of a cell can be equated to a highly efficient assembly line that produces many types of biological nanomachines. For example, in the translation of ribonucleic acid (RNA) to proteins, individual protein-units are assembled together into a ~ 30-nm sized functional system called a ribosome. The ribosome interacts with RNA and translates the RNA-sequence into an amino acid sequence. This amino acid sequence then interacts with other biological systems to fold into an active protein. Eukaryotic ribosomes contain up to 82 proteins that assemble in a precise fashion to form this RNA-translational machine. This is just one example. Other sample systems include the proteins involved in the transcription process, RNAsomes that are involved in splicing activities, or protein-systems involved in extracellular signalling. Although we have not reached or even come close to such sophistication in designing nanoscale-devices, biological machineries provide design guidelines and inspirations for building nanoscale devices.

Mimicking biological systems for designing nanoscale-devices may be a powerful strategy. The controlling factors in coordinating the 20-amino acids and 4-bases into functional units are non-covalent interactions. These interactions include hydrophobic-

hydrophobic interactions, van der Waals forces, hydrogen bonding, and molecular stacking. Nanoparticles, in general, are extremely dependent on molecular forces to assist them in maintaining their monodispersity. The ability to coordinate the molecular forces, similar to the ability of proteins to properly fold into a functional unit, on the surface of nanoparticles is difficult to do at this particular point since protein folding mechanisms remain somewhat unclear. A simpler and first step toward the use of biology to mediate nanoparticle assembly is the use of recognition biomolecules (RBs). RBs can be coated onto the surface of nanoparticles and be used to direct the formation of nanoparticle aggregates.

The simple mixing of antibody-coated nanoparticles with their matching antigen in aqueous solution can lead to rapid aggregation of nanoparticles.[16] The ordering of nanoparticles within the aggregate network can be coordinated by the RBs. For example, proteins such as avidin have four binding sites to the small organic molecule biotin as shown in Figure 2; antibodies are inhomogeneous in structure and have three binding sites. These two protein systems will yield nanoparticle-aggregates of different shape, size, and nanoparticle spacing. For nanoparticle assembly, proteins may one day act as "molecular glue" to join nanoparticles but will unlikely be used for building nanodevices that require dynamic assembly since protein-induced aggregation may be irreversible. Beyond network formation, the conjugation of proteins to the surface of nanoparticles provides nanoparticles with greater functional capabilities. Montemagno and coworkers proposed the powering of inorganic nanodevices with biomolecular motor.[12, 21] They demonstrated that biomolecular motors such as F_1-adenosine triphosphate synthase and myosin provide enough force to propel inorganic nanoparticles in solution. Vogel and coworkers proposed the use of motor proteins with microtubule track systems to construct molecular conveyer belts to build nanoscale devices.[22, 23] They are mimicking the biological process of vesicle transport inside cells.

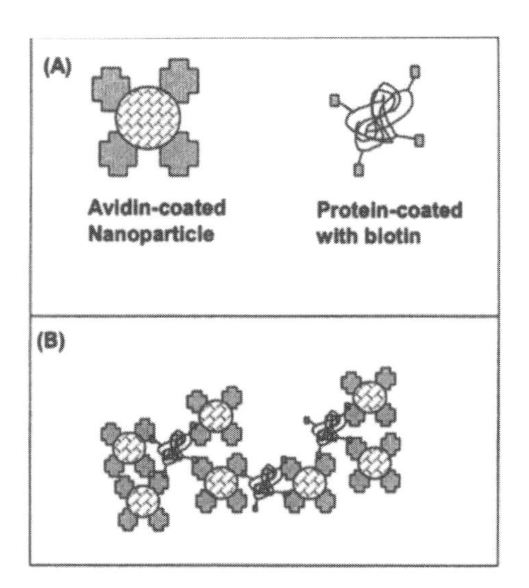

FIGURE 2. Schematic of proteins directing the assembly of nanoparticles. Avidin, a protein that binds to the vitamin biotin, is coated onto the surface of the nanoparticle. The addition of biotin-conjugated protein (e.g., albumin) to a solution of avidin-coated nanoparticle leads to aggregation due to the interactions between biotin and avidin.

Oligonucleotides, another biomolecule, provide great versatility in the assembly process; see Figure 3. Oligonucleotides are short-fragments of DNA or RNA that can be easily synthesized using a machine. Similar to a zipper, single-stranded oligonucleotide sequences can hybridize, or pair-up, with a matching sequence through hydrogen-bonding interactions. They dehybridize simply by heating. In 1996, Alivisatos and coworkers[24]and Mirkin and coworkers[25] were two of the first groups to describe the use of oligonucleotides to assemble nanoparticles. Alivisatos and coworkers demonstrated the selective spacing of gold nanoparticles on surfaces while Mirkin and coworkers demonstrated the reversible aggregation of gold nanoparticles in solution.

Thiolated-oligonucleotides (e.g., HS-ATGGCCA where A, T, G, C refers to the nucleic acid bases adenine, thymine, guanine, cytosine) are directly adsorbed onto the surface of gold nanoparticles and excess oligonucleotides are removed from solution using ultracentrifugation. A solution of ATGGCCA-coated gold nanoparticles can aggregate in the presence of the palindrome sequence TACCGGTTGGCCAT due to the formation of an ATGGCCA-TACCGGT duplex. Upon formation of the duplex structure, the solution changes colour from ruby red to blue. This indicates the assembly of gold nanoparticles into an aggregate. Heating of the solution beyond the melting temperature induces the dehybridization of the ATGGCCA-TACCGGT duplex, returning the solution colour back to red.

The properties of the oligonucleotide-induced aggregates can be easily manipulated. Changing the length of the oligonucleotide sequence can alter the length between nanoparticles in the aggregate network. Complex structures can also be developed – for example, Mucic et al. demonstrated the construction of binary nanoparticle networks where 8-nm gold nanoparticles were assembled onto the surface of a 31-nm gold nanoparticle as illustrated in Figure 3.[26] Different types of nanoparticles can be introduced into the oligonucleotide aggregates. Enzymes that can cleave specific DNA sequences (e.g., EcoRI) may provide another tool for producing unique structures in the aggregate network.[27, 28] Oligonucleotides provide a versatile "molecular glue" to organize nanostructures for engineering nanoscale devices. However, there are still great difficulties in producing aggregate structures with controlled size (e.g., structures with only 2, 3, or 4 particles) or shape (e.g., 3-D cubical vs. spherical shaped aggregate) using oligonucleotide-based methodologies.

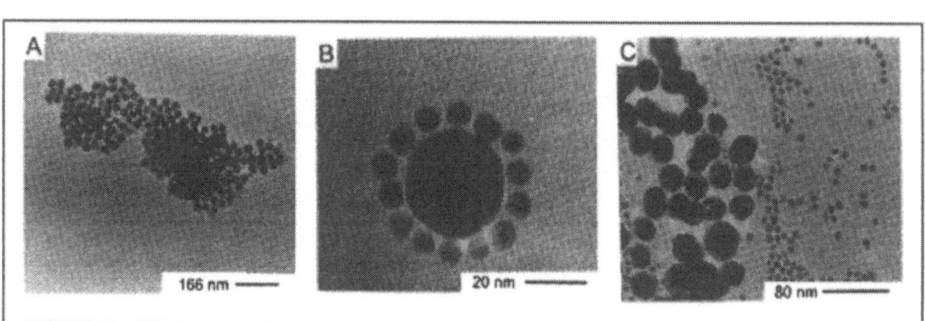

FIGURE 3. TEM images of nanoparticle assembly. 8-nm colloidal gold nanoparticles were coated with the sequence 3'HS$(CH_2)_3$O(O)OPO-ATG-CTC-AAC-TTC and 31-nm colloidal gold nanoparticles were coated with the sequence 3' TAG-GAC-TTA-CGC-OP(O)(O)O$(CH_2)_3$SH. (A) Upon addition of the linking sequence 5'TAC-GAG-TTG-AGA-ATC-CTG-AAT-GCG to a solution of the oligonucleotide-coated 8 and 31-nm gold nanoparticle, aggregation was observed. The 31-nm gold nanoparticles were covered with 8-nm gold nanoparticles. (B) By varying the ratio of 8-nm to 31-nm (in this example, it was 120:1), satellite structures can be formed. (C) When the linking sequence does not match either of the oligonucleotide sequences on the 8 or 31-nm gold nanoparticles, the nanoparticles do not aggregate. This figure is adapted from Ref. 26 with permission.

Surface-based techniques have recently been employed to provide a second level of control for nanoparticle aggregate formation. In a recent study by Bashir and co-workers, biotinylated oligonucleotide sequences were placed onto the surface of a silicon substrate.[29] Avidin-coated nanoparticles were then incubated with the substrate and were attached onto the surface through an avidin-biotin interaction. The addition of excess avidin-coated nanoparticles produced a monolayer of nanoparticles that can be released from the surface by dehybridizing the oligonucleotide sequence. Other surface-based approaches have utilized molecular templating as a strategy to organize nanoparticles. In one example, Stupps and coworkers developed nanometer-scaled ribbon structures out of polymers and then nucleated and grew semiconductor nanoparticles onto these ribbon structures.[30] Although the use of bio-inspired approaches toward nanoparticle assembly has provided a first level of control, this is nowhere near the level of complexity that cells have used to assemble their biomolecular machines. The next step in nanoscale-device building is to develop novel methods that can lead to a high level of control for nanoparticle assembly and that can integrate different-types of nanoparticles into a complex 3-D network, just like the ribosome. Once that has been achieved, we will be one-step closer to building smart and usable nanoscale devices.

4. MICROBIAL SYSTEMS FOR ASSEMBLING NANOSTRUCTURES

The use of microbial systems is another promising approach toward building nanoscale devices. Microbial systems may be used as templates for organizing nanoparticles or be programmed to express a set of worker proteins that can organize nanoparticles into functional devices.

Viruses possess hundreds of unique sizes and shapes that may be useful as templates for organizing nanoparticles. Emory and coworkers are proposing the development of optical sensing surfaces by coordinating SERS-active nanoparticles (SERS, surface-enhanced Raman spectroscopy) onto the surface of tobacco mosaic viruses (TMV).[31] The surface of the TMV is chemically modified to attract nanoparticles. For example, primary thiol atoms can be introduced onto the surface of the virus using Traut's reagent (a.k.a. iminothiolane), which modifies the surface of the virus to attract metallic nanoparticles. Metallic nanoparticles have a high binding affinity toward primary thiols. Therefore, the overall structures of the nanoaggregate are similar in shape and dimension to the virus. Furthermore, the ability to mutate the protein coating of virus may be useful for tuning the spacing between nanoparticles as well as for mediating nanoparticle heterogeneity on the surface of viruses.

Viruses can also be genetically engineered to produce a set of recognition peptides that can selectively identify nanoparticles based on their composition and lattice structure as illustrated in Figure 4. This technique is known as phage-display, where specific or unknown substrates, are panned against a library of phage particles to identify recognition molecules. With this technique, genes are inserted into viral particles called phage, and expressed as peptides on the outer perimeter of the phage after replication in bacteria. These genes can be mutated or varied to create a diverse pool of phage that can be screened against nanoparticles. Although, phage-display has been traditionally used in molecular biology, Belcher and coworkers pioneered the use of phage-display for nanotechnology research.[32, 33] They demonstrated the ability to program phage-particles to selectively identify specific type of nanoparticles and to use the phage as a means of organizing these nanoparticles into a 3-D layer. Phage-particles may one day be programmed to assist in the build-up of nanoscale devices.

Yeast cells also have the capability to build nanoscale devices. In 1989, Winge and coworkers discovered the ability of the yeast cells *Candida glabrata* or

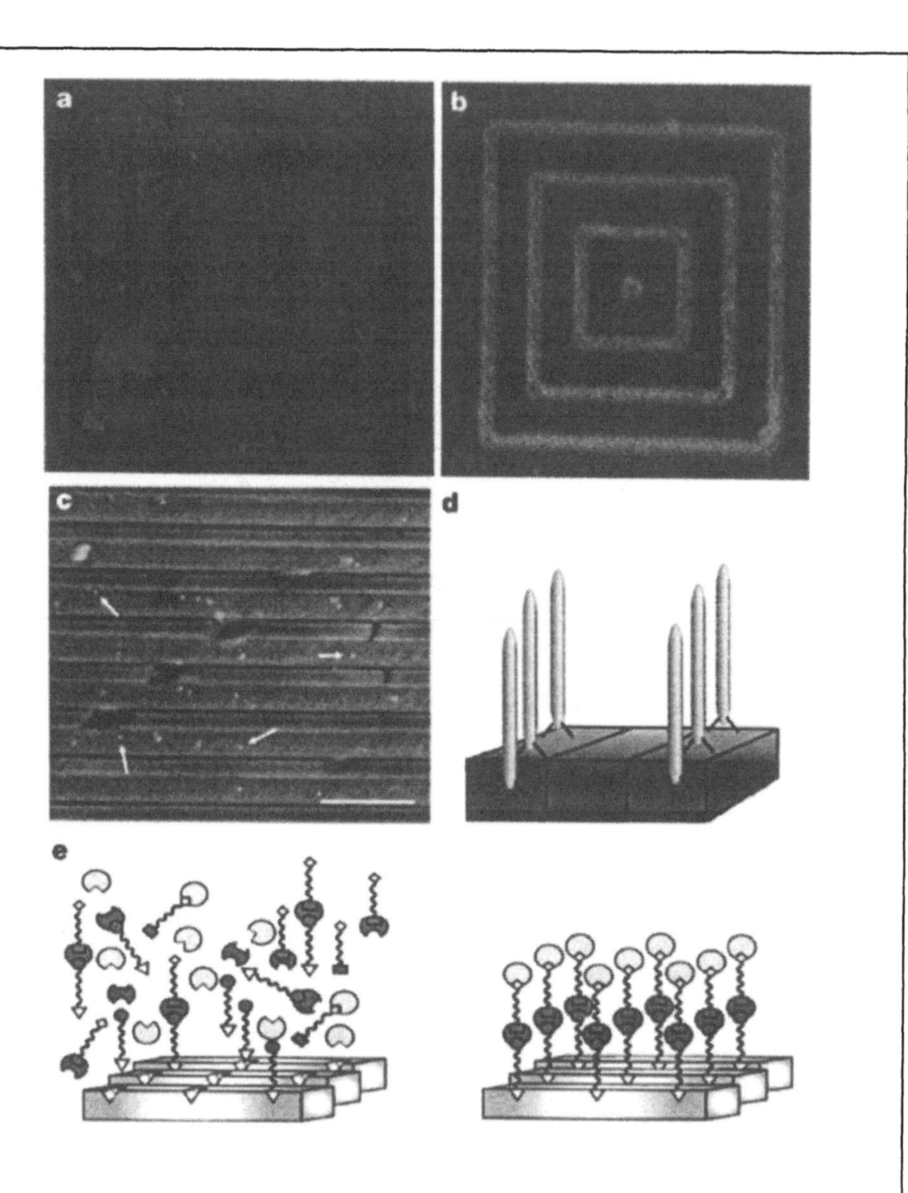

FIGURE 4. This figure depicts the use of phage-display screening to select for recognition molecules that can bind to inorganic targets. (A-B) An optical image of inorganic surface (1-μm GaAs and 4μm SiO_2). The autofluorescence of the SiO_2 was observed. Upon incubation with tetramethylrhodamine (TMR)-phage particles that recognized the GaAs surface, a high fluorescence emission was observed on the thinner lines. TMR is a fluorophore used to indicate the direct binding of phage particles to targeted site. (C) To further verify the selective binding of phage particles, the phage-particles were tagged with gold nanoparticles. The surface consisted of GaAs and $Al_{0.98}Ga_{0.02}As$. As shown in panel B, a SEM image revealed gold nanoparticles were stationed only on the GaAs region of the surface (shown by the arrows). Scale bar is 500 nm. (D) Diagram of the selection and binding process. (E) Diagram showing the use multiple phage-particles to direct assembly. This figure is adapted from Ref. 32 with permission.

Schizosacharomyces pombe to synthesize semiconductor nanocrystals.[34] Feeding yeast with Cd^{2+}- atoms can lead to the expression of a family of proteins that assist in the construction of ~ 3.0 nm CdS-nanoparticles. Once, the desired size of the nanoparticles is reached, yeast produces a glutathione-like protein to coat the surface of the nanoparticles. This protein helps to stabilize the nanoparticle from aggregation inside the cell and to direct the nanoparticles out of the cell. Recent studies have demonstrated the synthesis of different-types and sizes of nanoparticles using yeast cells.[35-37]

5. CURRENT STATE AND HIGHLIGHT OF BIOAPPLICATIONS OF NANOSTRUCTURES

Although the successful design of nanoscale devices is several decades away, nanoparticles have been integrated with biological molecules for numerous biomedical applications. These can be considered, in some manners, to be simple, monofunctional nanostructures that can manipulate the release of drug molecules or to detect biomolecules in solution.

One of the most interesting concepts in functional nanoscale systems is the integration of thermosensitive hydrogels with colloidal nanoparticles for optically controlling drug release. Halas and West described the use of gold nanoshells for localized drug delivery.[38] These nanoshells are made from a silica dielectric core with a gold shell, and their plasmon resonance wavelength can be tuned by varying the size of the core and the shell. It has been well known that colloidal gold nanoparticles and more recently, colloidal nanoshells have large absorption cross-sections (e.g., 5-nm diameter gold particle has a cross-section 3 nm^2 at 514 nm). Because of this, colloidal gold nanoparticles and nanoshells possess a photothermal effect that is induced by non-radiative collisions between optically excited electrons with solvent molecules. Thermal-sensitive polymers such as the copolymer N-isopropylacrylamide and acrylamide can undergo a phase-change when the polymer is heated above their lower critical solution temperature (LCST). At a temperature > LCST, molecules trapped inside the polymer are released into the external environment due to the shrinking of the polymer. Essentially, molecules such as drug agents that are trapped in the core are squeezed out of the hydrogel when heated. The development of thermal-sensitive polymer matrices on the surface of gold nanoshells can be utilized for storage of drug agents while the gold nanoshells can act as an optical switch to control the release of drug agents. It is believed that one day nanoparticle/drug storing system can be directly targeted to lesions *in vivo*. The advantage is that the drug molecules are protected from the immune defence systems *in vivo*; and can be selectively released at sites of injury - which is expected to reduce side effects.

Recently, Ruoslahti and coworkers have demonstrated the successful targeting of nanoparticles to specific sites in living animals.[14] They discovered peptides that specifically target tumour vasculatures.[39, 40] Two of these peptides were conjugated onto the surface of two different-emitting quantum dots. These peptide-conjugated quantum dots were introduced into a mouse bearing xenograft tumours through injection into the tail vein. After 20 minutes, two distinct fluorescence signals were distinctly apparent on the tumour tissues upon optical excitation. Co-staining with markers showed the localization of red-emitting quantum dots in the blood vessels while green-emitting quantum dots localized in the lymphatic vessels. No significant quantum dot emission was apparent in other nearby tissues and organs. Targeting molecules have been identified for tumour vessels and normal vessels in the brain, kidney, lungs, skin, pancreas, and other tissues.[41] The integration of targeting molecules with nanostructures should provide a means of delivering nanostructures to specific cells and tissues *in vivo* as well as nanoparticle-based contrast agents for ultrasensitive optical imaging.

Beyond *in vivo* applications of nanostructures, interfacing organic molecules with inorganic nanostructures have led to the development of a new generation of *in vitro* detection systems. Mirkin and coworkers harnessed the unique absorption characteristics of aggregated and non-aggregated gold nanoparticles for the detection of genetic mutations.[42, 43] Natan and coworkers developed a metallic-barcoding system that can analyze thousands of biomolecules simultaneously.[44] Colloidal metallic nanoparticles have also been utilized for genomic and proteomic screening. Dai and coworkers developed highly selective electronic biosensors by using protein-adsorbed carbon nanotubes.[45] Furthermore, there has been significant progress in the development of semiconductor quantum dots for biosensing and detection applications.[46] Other types of nanoparticles such as fullerenes and dendrimers have found applications in drug storage and delivery. Magnetic nanoparticles are utilized as contrast agents for enhancing MRI-imaging. Monofunctional nanostructures are rapidly advancing toward everyday use in research labs. In the future, we foresee the development of multifunctional nanostructures where onset and evolution of a disease can be sensed by the nanostructure; the sensing of the disease, then, cause the selective release of one type or a combination of drug agents.

6. CONCLUSIONS

The field of nanotechnology has great potential to change the world. There has been a tremendous focus in the last thirty years on developing and characterizing nanostructure materials. Nowadays, the goal is to utilize these materials as precursors to build nanoscale devices and to develop novel approaches to assemble these precursor nanostructurs into a functional device. Biology offers an excellent guide for assembling nanostructures since a cell can produce thousands of different functional units with only 20-different amino acid building blocks. Biomolecules such as proteins, oligonucleotides and microbial systems have been successfully applied toward organizing nanostructures into macrostructures. Although we have not built complex and functional nanostructures, there are numerous examples in the literature that demonstrate the utility of simple monofunctional nanostructures for biosensing and imaging applications, and drug storage/release systems. In the future, the ability to assemble nanostructures into complex functional units should produce novel systems that will have a broad and significant impact.

7. ACKNOWLEDGEMENTS

S. M. would like to thank the National Sciences and Engineering Resource Council of Canada (NSERC) for graduate fellowship. W. J. would like to thank the Ministry of Training, Colleges and Universities for the Ontario Graduate Scholarship. W. C. W. C. would like to thank the University of Toronto, Connaught Foundation, Canadian Foundation for Innovation, and Ontario Innovation Trust for research support.

8. REFERENCES

1. G. Frens, Controlled nucleation for the regulation of the particle size in monodisperse gold suspensions, *Nature Physical Science* **241** 20-22 (1973).
2. R.H. Baughman, A.A. Zakhidov, and W.A. De Heer, Carbon nanotubes - the route toward applications, *Science* **297** 787-792 (2002).
3. M.A. Hines and P. Guyot-Sionnest, Synthesis and characterization of strongly luminescing zns-capped cdse nanocrystals, *J. Phys. Chem. B* **100** 468-471 (1996).
4. M.A. Hines and P. Guyot-Sionnest, Bright uv-blue luminescent colloidal znse nanocrystals, *J. Phys. Chem. B* **102** (19), (1998).
5. C.B. Murray, D.J. Norris, and M.G. Bawendi, Synthesis and characterization of nearly monodisperse cde (e = s, se, te) semiconductor nanocrystallites, *J. Am. Chem. Soc.* **115** 8706-8715 (1993).
6. S. Park, Kim, S., Lee, S., Kim, Z. G., Char, K., Hyeon, T., Synthesis and magnetic studies of uniform iron nanorods and nanospheres, *Journal of the American Chemical Society* **122** 8581-8582 (2000).
7. L. Manna, D.J. Milliron, A. Meisel, E.C. Scher, and A.P. Alivisatos, Controlled growth of tetrapod-branched inorganic nanocrystals, *Nature Materials* **2** (6), 382-385 (2003).
8. B. Nikoobakht, El-Sayed, M. A., Preparation and growth mechanism of gold nanorods (nrs) using seed-mediated growth method, *Chemistry of Materials* **15** 1957-1962 (2003).
9. B.M.I. Van Der Zande, M.R. Bohmer, L.G.J. Fokkink, and C. Schonenberger, Colloidal dispersions of gold rods: Synthesis and optical properties, *Langmuir* **16** (2), 451-458 (2000).
10. G.P. Mitchell, Mirkin, C. A., Letsinger, R. L., Programmed assembly of DNA functionalized quantum dots, *Journal of the American Chemical Society* **121** 8122-8123 (1999).
11. R. Jin, G. Wu, Z. Li, C.A. Mirkin, and G.C. Schatz, What controls the melting properties of DNA-linked gold nanoparticle assemblies?, *Journal of the American Chemical Society* **125** (6), 1643-54 (2003).
12. R.K. Soong, Bachand, G. D., Neves, H. P., Olkhovets, A. G., Craighead, H. G., Montemagno, C. D., Powering an inorganic nanodevice with a biomolecular motor, *Science* **290** 1555-1558 (2000).
13. C. Montemagno and G. Bachand, Constructing nanomechanical devices powered by biomolecular motors, *Nanotechnology* **10** (3), 225-231 (1999).
14. M.E. Akerman, W.C. Chan, P. Laakkonen, S.N. Bhatia, and E. Ruoslahti, Nanocrystal targeting in vivo, *Proceedings of the National Academy of Sciences of the United States of America* **99** (20), 12617-21 (2002).
15. Y. Kang, Taton, T. A., Micelle-encapsulated carbon nanotubes: A route to nanotube composites, *Journal of the American Chemical Society* **125** 5650-5651 (2003).
16. W.C. Chan and S. Nie, Quantum dot bioconjugates for ultrasensitive nonisotopic detection.[comment], *Science* **281** (5385), 2016-8 (1998).
17. R. Chakrabarti and A.M. Klibanov, Nanocrystals modified with peptide nucleic acids (pnas) for selective self-assembly and DNA detection, *J. Am. Chem. Soc.* **125** 12531-12540 (2003).
18. K. Kamaras, Itkis, M. E., Hu, H., Zhao, B., Haddon, R. C., Covalent bond formation to a carbon nanotube metal, *Science* **301** 1501 (2003).
19. M. Bruchez, Jr., M. Moronne, P. Gin, S. Weiss, and A.P. Alivisatos, Semiconductor nanocrystals as fluorescent biological labels., *Science* **281** (5385), 2013-6 (1998).
20. X.E.A. Wu, Immunofluorescent labeling of cancer marker her2 and other cellular targets with semiconductor quantum dots, *Nature Biotechnology* **21** 41-46 (2003).
21. C. Montemagno, Biomolecular motors: Engines for nanofabricated systems., *Abstracts of Papers of the American Chemical Society* **221** U561-U561 (2001).
22. H. Hess, J. Clemmens, J. Howard, and V. Vogel, Surface imaging by self-propelled nanoscale probes, *Nano Letters* **2** (2), 113-116 (2002).
23. H. Hess, J. Clemmens, D. Qin, J. Howard, and V. Vogel, Light-controlled molecular shuttles made from motor proteins carrying cargo on engineered surfaces, *Nano Letters* **1** (5), 235-239 (2001).
24. A.P. Alivisatos, Peng, X., Wilson, T. E., Johnsson, K. P., Loweth, C. J., Bruchez, M. P., Schultz, P. G., Organization of nanocrystal molecules using DNA, *Nature* **382** 609-611 (1996).
25. C.A. Mirkin, Letsinger, R. L., Mucic, R. C., Storhoff, J. J., A DNA-based method for rationally assembling nanoparticles into macroscopic materials, *Nature* **382** 607-609 (1996).
26. R.C. Mucic, J.J. Storhoff, C.A. Mirkin, and R.L. Letsinger, DNA-directed synthesis of binary nanoparticle network materials, *Journal of the American Chemical Society* **120** (48), 12674-12675 (1998).
27. J.R. Taylor, Fang, M. M., Nie, S., Probing specific sequences on single DNA molecules with bioconjugated fluorescent nanoparticles, *Analytical Chemistry* **72** 1979-1986 (2000).

28. J. Liu, Lu, Y., A colorimetric lead biosensor using dnazyme-directed assembly of gold nanopaticles, *Journal of the American Chemical Society* **125** 6643-6643 (2003).
29. H. Mcnally, Pingle, M., Lee, S. W., Guo, D., Bergstrom, D. E., Bahir, R., Self-assembly of micro- and nano-scale particles using bio-inspired events, *Applied Surface Science* **214** 109-119 (2003).
30. E.D. Sone, Zubarev, E. R., Stupps, S. I., Semiconductor nanohelices templated by supramolecular ribbons, *Angew. Chem. Int. Ed.* **41** 1705-1709 (2002).
31. S.R. Emory, http://www.chem.wwu.edu/dept/facstaff/semory/semoryresearch.shtml).
32. S.R. Whaley, English, D., Hu, E., Barbara, P. F., Belcher, A. M., Selection of peptides with semiconductor binding specificity for directed nanoparticle assembly, *Nature* **405** 665-668 (2000).
33. S. Lee, Mao, C., Flynn, C., Belcher, A. M., Ordering of quantum dots using genetically engineered viruses, *Science* **296** 892-895 (2002).
34. C. Dameron, R.N. Reese, R.K. Mehra, A.R. Kortan, P.J. Carroll, M. Steigerwald, L. Brus, and D.R. Winger, Biosynthesis of cadmium sulfide quantum semiconductor crystallites, *Nature* **338** 596-597 (1989).
35. W. Bae, Abdullah, R., Henderson, D., Mehra, R. K., Characteristics of glutathione-capped zns nanocrystallites, *Biochemical Biophysical Research Communications* **237** 16-23 (1997).
36. W. Bae, Mehra, R. K., Properties of glutathione- and phytochelatin-capped cds bionanocrystallites, *Journal of Inorganic Biochemistry* **69** 33-43 (1998).
37. W. Bae, Mehra, R. K., Cysteine-capped zns nanocrystallites: Preparation and characterization, *Journal of Inorganic Biochemistry* **70** 125-135 (1998).
38. S.R. Sershen, S.L. Westcott, N.J. Halas, and J.L. West, Temperature-sensitive polymer-nanoshell composites for photothermally modulated drug delivery, *Journal of Biomedical Materials Research* **51** (3), 293-298 (2000).
39. P. Laakkonen, K. Porkka, J.A. Hoffman, and E. Ruoslahti, A tumor-homing peptide with a targeting specificity related to lymphatic vessels, *Nature Medicine* **8** (7), 751-5 (2002).
40. K. Porkka, P. Laakkonen, J.A. Hoffman, M. Bernasconi, and E. Ruoslahti, A fragment of the hmgn2 protein homes to the nuclei of tumor cells and tumor endothelial cells in vivo, *Proceedings of the National Academy of Sciences of the United States of America* **99** (11), 7444-9 (2002).
41. E. Ruoslahti, Specialization of tumor vasculature, *Nature Reviews Cancer* **21** 84-90 (2002).
42. R. Elghanian, J.J. Storhoff, R.C. Mucic, R.L. Letsinger, and C.A. Mirkin, Selective colorimetric detection of polynucleotides based on the distance-dependent optical properties of gold nanoparticles.[comment], *Science* **277** (5329), 1078-81 (1997).
43. R.L. Letsinger, C.A. Mirkin, S.J. Park, G. Viswanadham, and L. Zhang, Poly(oligonucleotide) conjugates: Applications in assembling nanoparticles and in detecting DNA sequences, *Nucleic Acids Research. Supplement* (1), 1-2 (2001).
44. S. Nicewarner-Pena, R.G. Freeman, B.D. Reiss, L. He, D.J. Pena, I.D. Walton, R. Cromer, C.D. Keating, M.J. Natan, Submicrometer metallic barcodes, *Science* **294** 137-141 (2001).
45. R.J. Chen, S. Bansaruntip, K.A. Drouvalakis, N.W.S. Kam, M. Shim, Y.M. Li, W. Kim, P.J. Utz, H.J. Dai, Noncovalent functionalization of carbon nanotubes for highly specific electronic biosensors, *Proceedings of the National Academy of Science* **100** 4984-4989 (2003).
46. W.C. Chan, D.J. Maxwell, X. Gao, R.E. Bailey, M. Han, and S. Nie, Luminescent quantum dots for multiplexed biological detection and imaging, *Current Opinion in Biotechnology* **13** (1), 40-6 (2002).

BRIDGING NATURAL NANO-TUBES WITH DESIGNED NANOTUBES

Duan P. Chen*

1. INTRODUCTION

Nanotechnology is concerned with developing nanometer systems at the atomic (molecular or macromolecular) level that have novel properties and functions because of their intrinsic nanometers (nm) length scale. At this small size scale (under 100 nm), a system may exhibit its novel differentiating properties and functions. Nanotechnology utilizes the novel properties of these engineered small systems (alone or integrated with other systems) to make new discoveries and to contribute to the development of new technologies.

On the other hand, nature has already utilized nano-length-scale systems to perform crucial biological functions. These critical biological functions require high specificity and efficiency. As a result of the utilization of intrinsic molecular interactions at the nanoscale, natural biological systems achieve such specialized functions, especially those demonstrated in highly-specialized proteins. For example, ionic channel proteins in the membranes of cells are key biological elements responsible for a variety of signaling processes that control a medley of functions from contraction to secretion.[1] They are a few nanometers in length and a few angstroms to nanometers in diameter, yet many diseases are associated with ion channel defects or mal-functions.[2, 3] Not only are their spatial dimensions measured in nanometers, but also they operate in the nanosecond time regime.

This chapter highlights potential applications of nature biological systems in combination with the engineered carbon nanotubes in device applications, sensing, drug-delivery, and computations. An analogy of ion channels to semiconductor devices is made and elucidated.

*Department of Molecular Biophysics & Physiology, Rush University Medical Center, Chicago, Illinois 60612. Email: dchen@rush.edu

2. ION CHANNELS: WHAT THEY ARE AND HOW THEY ARE STUDIED

Ion channels are ion-conducting membrane proteins, and their aqueous pathway can be opened (gated) by ligands or voltage changes (To be exhaustive: Some can be opened by second messengers like G-proteins. Others open their pathways by responding to external mechanic pressures induced by membrane curvatures. Some others are believed to be open all the time like gap junctions and porins). The ionic electrical signals, facilitated by ion channels, are then used to control biological functions throughout living systems. Membranes (phospholipid bilayers) form a low dielectric barrier to hydrophilic and charged ions (or molecules), insulating the cell interior from the exterior electrically.

Once open, its water-filled pore is highly selective for a specific ion species as in most voltage-gated channels, or non-selective for cations or anions as in most ligand-gated channels. The structures of the pore (especially at its selectivity filter region) and its gates (the regions control the opening/closing of the pathway) are highly conserved in evolution.

Ion channels are highly functionalized complex proteins. Individual ion channel protein generates all-or-nothing cellular electric signals carried by ions with high selectivity; therefore, their functions require a highly complex protein structure. They are usually composed of multiple subunits (4 to 6) and each subunit can have many transmembrane segments. The segments of ion-channel consist of amino acids in a linear fashion. But the final three-dimensional molecular folding structure determines its eventual functions, such as specificity of bindings and sensitivity to external voltage. Traditionally ion channels are pictured to have the four functional parts: ion-conducting pore (the ionic pathway), selectivity filter (the filter determines what kind of ion can pass through), gates (open and close channel pathway, like a door), and voltage sensors or binding sites (characteristics of the channels open by voltage or by endogenous and exogenous ligands). With molecular biology, the functional parts of these four components are gradually realized in their linear amino acid sequences. The precise structure-function relation of those key four parts remains a challenge as few ion channel three-dimensional structures are known.

The structure of porin is the first ion channel structure solved at atomic resolution at a 3 Å x-rays resolution.[4, 5] Porins are a family of proteins that form channels across the outer membranes of Gram-negative bacteria. The wild-type OmpF porin is a trimer as shown in Fig. 1. In Fig. 1, the OmpF structure[4] is graphed with RasMol computer program[6]. Porin family is of homo-trimeric channel proteins. The wild-type porin OmpF (from outer membrane of Escherichia coli) is surrounded by a 16-stranded antiparallel $\beta-$barrel, forming an trimeric ion pathway. The $\beta-$strands are amphipathic, as they contain alternating polar and non-polar residues. The antiparallel inter-strand interaction is fully facilitated by saturated H-bonds. This creates a hydrophilic interior providing a water filled channel. This pathway is of hour-glass shape. The β-barrels have one loop extending towards the pore region, constricting the narrowest region to have a radius of 5 Å. The narrowest pore region channel has a cross section of 0.8x1.1 nm. It is so large that it allows polar permeants up to a molecular weight of 600 Dalton.[7] This large pore is therefore non-selective for small ions. Even though the pore of OmpF porins is relatively wide (not much wider than other natural channels, for example, see Refs. 8 and 9), the actual permeability is complicated[10, 11, 12] and important[13] as well.

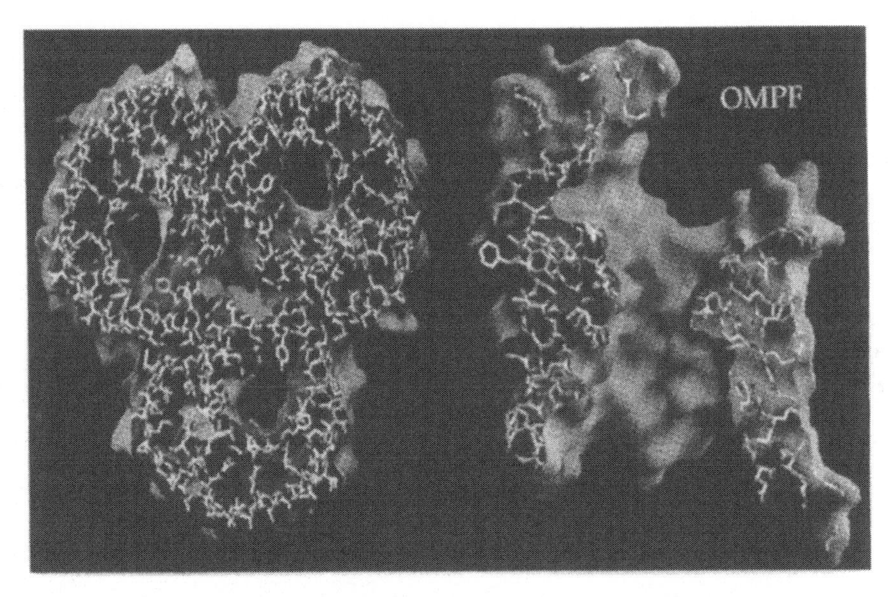

Figure 1. The structure of a wild-type OmpF porin channel. On the left is the ion conducting trimmer, and on the right is a view of the pore region of a monomer. Pictures are generated by RasMol from its PDB structure.

The first crystal structure of ion selective channels is determined from the bacteria Streptomyces lividans (KcsA K^+ channel) with 3.2 Å resolution, shown in Fig. 2. (The actual channel structure from residue position 23 to 119 is determined and amino acids 126-158 at the carboxyl terminal end have been cleaved off.[14]) The x-rays structure reveals an inverted cone shaped channel of an overall length of 45 Å, with a four-fold tetrameric symmetry. Like several other membrane proteins, it has two rings of aromatic amino acids positioned to extend into the lipid bilayer, near the membrane-water interfaces. The channel cross section varies along the pathway with a narrowest region to be the selectivity filter, where the signature amino acid residues of K channel reside. The inner pore helices are tilted with respect to the membrane normal by about 25°, forming a cone-shaped pore opening with the wider part facing the outside of the cell. In this region is of 12 Å long, the carbonyl oxygens face the pore lining to form a negatively charged oxygen ring. This ring coordinates a dehydrated potassium ion, yet a sodium ion is too small to facilitate a strong interaction with this region, yielding the origin of ion-selectivity.

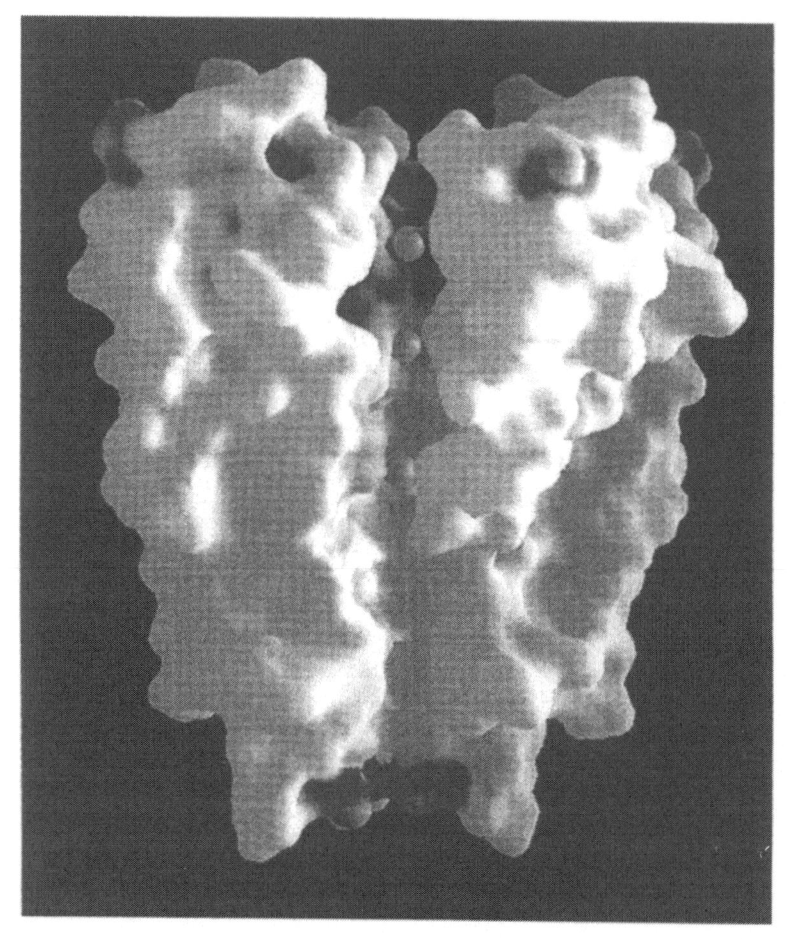

Figure 2. The structure of a potassium ion selective channel, KcsA channel. At the top is the selectivity filter region, where there are two potassium ion binding sites, denoted by green circles. Another binding site is at the cavity region below. The picture is taken from Doyle 1998.

At the other end, it is a tunnel of 18 Å, that leads to a wide cavity of 10 Å at the middle of the channel. The entire channel is predominantly hydrophobic except at the selectivity filter region and at the cavity. At these places, the crystal structure with permeating ions shows three ion binding sites. One site is stabilized in the middle of the membrane within an aqueous cavity with four the negatively charged carboxyl ends (helix dipole, one from each monomer) of four central α-helices. Because of the size of the cavity, this central ion is proposed to be hydrated. The other two sites near each other are within the selectivity filter. Two potassium ions at close proximity within the selectivity filter repel each other so that the energetic barrier of permeation is significantly reduced. Furthermore, the overall pore lining is mainly hydrophobic. Together, it might explain why most K-channels have a very high flux rate with an extremely high selectivity of potassium over sodium ions (four orders of magnitude in conductance).

Experimentally, individual ion channels are generally studies by patch-clamp technique with the living cells or by bilayer reconstitution technique when there are purified ion channel proteins available.[15] The all or none fashion electrical currents of individual ion channels are in the magnitude of pico Amperes. They open in a fraction of seconds spontaneously; however, the statistical average of the mean open time (probability) varies with the externally applied voltage. Transitions between open and close states are very fast and in the order of fractions of a millisecond, and appear in the recordings as rectangular jumps from one level to the other, like a random telegraph signal. Within each open event, there are millions of ions pass through each ion channel.

Theoretically, ion permeation is traditionally studied by the classical barrier models. Typical barrier models have difficulties fitting a wide range of experiments[1] especially at large applied transmembrane voltages. A direct comparison of electrodiffusion model with the barrier model has shown that barrier models can give accurate electrodiffusion descriptions when barriers are high and far from boundaries.[16, 17] Alternatively, molecular dynamic studies[18, 19, 20, 21] are a wonderful tool to analyze local interactions, to sieve out bad (high-energy) interactions, and to investigate the stability of structure near. However, they have the advantage of investigating possible ion-ion interaction and correlation effects.[22] If combined with other approaches, such as, brownian dynamics (or Langevian dynamics) and continuum electrodiffusion, molecular dynamics can be a powerful tool to verify some of the microscopic assumptions used in those approaches. The recent study, using molecular dynamics simulation and brownian dynamics, has verified the applicability and accuracy of continuum electrodiffusion theory for wild-type OmpF.[23] The high cation occupancy found in the molecular dynamics simulations even at extremely low ionic solution (as low as 1μM KCl) is the strong electrostatic buffering properties of a charged ion channel pore, reported in the study using the continuum electrodiffusion description.[16, 24, 25]

In the continuum electrodiffusion description, the couple are integrated to calculate the flux of ions (under diffusion and the flow driven by electric-field). In fact, it has been shown that the couple Poisson-Nernst-Planck (PNP) Equations can be derived by averaging ion Langevian motions,[17, 26] under the frequent collision and over-damped limit. The result shows that the concentration used in the continuum electrodiffusion description is the unconditioned probability with proper normalization to the number of particles. The result also shows that the effective mobile ions charge density is the conditioned probability distribution in the Poisson equation in the coupled system. As the conditional densities are closely related to the pair-correlation functions, the coupled PNP system has implicitly contain the ion-ion correlation (same as the pair-correlation functions) criticized in many studies.[26]

Interestingly, the electrodiffusion description of ion channel permeation by PNP Equations makes the ion channel as molecular semiconductor devices, because the

PNP equations are the semiconductor device equations — the drift-diffusion model. Apparently, the ion channels work at very different domain from the modern semiconductor devices. An analogy is shown in Table I. To incorporate other effects into the electrodiffusion description, such as the ion finite size effect, remains to be the subject of future studies.

3. POSSIBLE APPLICATIONS OF ION CHANNELS WITH CARBON NANOTUBES

The molecular recognition and detection require discovery and utilization of how nature performs these tasks in biology. The recognition and detection of molecules by biological systems is impressive, exceeding our technological capabilities in many respects, as described above in the case of selective potassium channel. Natural biological systems offer selectivity, sensitivity and efficiency, and yet the engineered carbon nanotubes give great controllability. It is possible and beneficial to marry the natural biological molecular recognition process by ion channel proteins with the cutting edge nanotechnology. In this way, a new hybrid devices/material can be create to possess unprecedented phenomena and properties. Specifically, the new hybrid devices will exhibit remarkable sensitivity, selectivity and efficiency that enable direct, real-time conversion of bio-molecular signals into electrical signals.

It will be possible to develop a new organic/inorganic based activity (electric, optical, chemical, mechanical, thermal or magnetic) detection system, which could have significant impact in many areas including environmental monitoring, disease surveillance, medical diagnostics, and development of therapeutics.

For example, ionic channel proteins (cylinder shaped, a few angstroms in diameter and tens of angstroms in length, as described above), found in lipid membranes of living cells, control the rapid passage of ions and the resulting current of charge. Ionic current is used throughout living systems to perform key physiological functions, such as, signaling in the nervous system, signal transduction in sensory organs, and coordination of muscle contraction. Channels help control transport in nearly every tissue and cell. Ion channels can be viewed as natural electrical transducers of chemical signals.

Ion channels are known for their selectivity: they pass material selectively. This unique sieving property is utilized by nature to do its biological recognition process. It has been demonstrated that ion channels can be used for sensing chemical compounds as well as virus, warfare pathogens and other small proteins. Ion channels are therefore an example of natural biological transducer system with highly desired sensitivity, specificity and efficiency for converting electrical input to an output signal, in additional to signals in other forms (chemical signals can be transduced by ligand sensitive channels).

However, biological systems like ion channels often do not have known structure and they can only be controlled on the average in statistical sense. But the engineered carbon nanotubes offer the controllability lacked in natural biological systems.

Carbon nanotubes, on the other hand, are similar in size as natural biological ion channels. The advantages of carbon nanotubes over natural ion channels are: 1) they have known structures, 2) they can be engineered into different geometry and shapes, 3) they have a high mechanical strength, which allows mechanical manipulation by other tools, such as atomic force microscopy, 4) they can be fully controlled deterministically; 5) they can be switched on and off between different conducting states.

It is then possible to take advantage of the selective properties of natural ion channels and the controllability of engineered nanotubes to yield an unique property, which neither system alone can achieve. Marrying ion channels with nanotubes for biological and medicinal application is the main theme of this chapter.

Utilizing the selective properties of ion channels, it is possible to detect a very low dose of a wide range of agents of interest including unknown, engineered, and emerging threat agents. Ion channels stand alone as proteins, and they do not require cells or tissues. Different ion channels can be developed to form a library of transducers responding to a broad spectrum of responses of interest. A nanotube, on the other hand, is a chemically stable, high-strength material, which does NOT have the inherent instability to extremes of temperature, hydration, ions, pH, and foreign materials. Furthermore, both ion channels (some of them) and nanotubes have defined structures, which enable computational design and analysis. System modeling can be performed to combine their unique individual properties in an integrated hybrid detection system. Simulations can guide the engineering and integration of a such hybrid system. Therefore, the potential applications are in the following areas:

1. Constructing bio-fluidic nano-plumbing devices

It is possible to construct nano-plumbing devices for transport in biological fluidic environment. It is expected that the nanotube is an ideal miniature device for biological fluidic transport. After all, a nanotube is a miniature of a glass tube (with a definite structure), which is used to pass aqueous solution daily. A nanotube is expected to be an ideal nano-plumbing device based on the observation that the walls of nanotubes have very similar dimension and chemical linings of typical ion channel proteins selected by nature to facilitate bio-fluidic transport. Furthermore, a nanotube is the high-strength yet light-weight pipe and possesses the mechanical strength to sustain the applied hydrodynamic pressure to facilitate a controlled biological fluidic transport. When necessary, the inner wall lining of a carbon nanotube can be changed chemically to adapt a more hydrophilic environment to facilitate the occupancy of water molecules and their transport through it. It can be tested if an engineered nanotube (including its chemical modifications) is a better biological fluidic transport apparatus. It will be shown whether an engineered nanotube is a better ion transport device across cell membrane because of its controllability. Will be identified are the key controlling factors in biological transport including the geometry of the nanotubes, the wall lining of the nanotubes, or other factors.

The specific technical challenge is: How to insert the nanotubes into a lipid membrane to make a pathway for biologic fluidic transport. We think that we can approach this challenge in three ways.

1) To use a speed controlled magnetic stirrer in the conventional electro-

physiological bilayer reconstitution experiment. In the study of ion channels with a bilayer setup, the purified ion channels are reconstituted across the lipid membrane by dissolving the ion channel proteins in the electrolyte solution. The reconstitution of ion channels is then enhanced mechanically by a small spinning magnetic stirrer in the bath where purified channel proteins are added. The necessary kinetic energy for ion channel insertion into the lipid membrane is then provided. The lipid membrane is first painted mechanically by hand onto an insulating plastic sheet with a micro size hole.

2) To use atomic force microscopy (AFM) as a tool to mechanically insert a nanotube into the lipid membrane. It is possible to pick up a nanotube and physically insert it across the lipid membrane because of its mechanical strength. This approach has an advantage over the first one: because the system can be controlled so that it is sure that one and only one nanotube will be inserted.

3) There is a possible third approach as well. One can attempt to coat the nanotube with the lipid membrane material. The exterior of the nanotubes has been functionalized by many researchers with organic, inorganic and protein structures.

Another challenge is how to keep the two ends of carbon nanotubes open. It can be handled by many methods, for example, the carbon nanotube can be dispersed in solvent and then ultrasonicated. A different approach is to zap the nanotube with a high voltage before it is inserted. A high voltage is known to open up the ends of a carbon nanotube.

One more challenge is how to make the hydrophilic modifications of carbon nanotubes. It can be accomplished by (1) physical adsorption of surfactants, (2) chemical modification with hydrophilic molecules; and (3) chemical attachment of polar groups at the two ends of a carbon nanotube. Furthermore, the properties of carbon nanotubes can be modified by the attachment of proteins or other biological molecules from the outside wall of a carbon nanotube, using the hydroxyl and amine chemistry.

Once a nanotube is inserted in a lipid membrane, it is expected that the lipid membrane will provide an extremely good insulating wall. Furthermore, it is known in biological patch-clamp technrique that lipid will form a tight seal around the outer wall of a nanotube, with a resistance as high as giga-Ohms, yielding a high throughput pathway along the nanotube.

The advantages of bio-fluidic nano-plumbing devices are:

- Facilitating CONTROLLED cellular transport by interconnecting two cells at the two ends of a nanotube in a dumb-bell shape
- Facilitating CONTROLLED cellular signal transduction
- Facilitating CONTROLLED delivery of a drug to targeted cells
- Possessing various geometrical shapes and that have high mechanical strength at the same time
- Possessing the ability to switch: a nanotube can be made to switch on and off electronically and deterministically

Therefore, it is then a controllable and switchable nanotube miniature device for biological fluidic transport and drug delivery.

2. Constructing ion channel/nanotube hybrid devices

It is possible to construct ion channel/nanotube hybrid devices by the manipulation of chemical bonds to connect ion channels and nanotubes using the atomic force microscope as the manipulating tool. The specific technical challenge is whether AFM can change a chemical bond. It has been recently demonstrated that atomic force microscopy can be used to manipulate every stage of a chemical reaction chain. At each step, a specific chemical bond is changed to yield a new intermediate product, until the final chemical product is produced. Therefore, it should be possible to alter a few chemical bonds to interconnect an ion channel with a nanotube.

Challenge: How to interconnect a biological ion channel with a carbon nanotubue; how to make ions go through carbon nanotubes.

Approach: AFM manipulation of chemical bonds; surface treatment and/or electrophoresis enhancement; attachment of molecules to the walls of carbon nanotube from outside.

3. Constructing ion channel based high capacity memory devices controlled by nanotubes

It is possible to construct a biological computational memory device based on ion channels with their gating controlled deterministically by a nanotube device. Ion channels work naturally in on-off (conducting and non-conducting) states stochastically. Therefore, they are natural binary computing devices. Furthermore, ion channels are known to aggregate together in high densities. Therefore, it is possible to employ them as the basic unit of high capacity computational memory devices, provided we can control them to turn on and off deterministically. On the other hand, one has total control of switching properties of a nanotube deterministically. Once ion channels are inter-connected with a nanotube device, one would have greater control of the hybrid device by controlling the switching of the nanotube. The underlying principle used to switch a nanotube can be applied to control ion channel opening and closing (gating).

The technical challenges are:

- To lock the ion channel in the open or close state
- To read out the memory, i.e., to test whether an ion channel is closed or open
- To use the switch of nanotube to control ion channel switch

The first challenge is likely be approached with the aid of a controlled nanotube. The second challenge can be approached by the patch-clamp technique to measure single channel activities. Furthermore, the glass pipette used in the conventional patch-clamp technique can be replaced with a specific shaped nanotube (with appropriate size). There are many advantages of a nanotube over a conventional glass pipette. For example, a nanotube has a defined geometry and known dielectric properties. The capacitance and dielectric property of a nanotube can be altered by engineering. For example, the dimension and geometry of a nanotube can be changed. Moreover, the number of layers on the wall of a nanotube (the thickness of the wall of a nanotube) can be changed as well.

4. Simulating bio-fluidic transport through nanotubes & through ion channel/nanotube hybrid devices

It is possible to study bio-fluidic transport through nanotubes by application of the Navier-Stokes equations. The phenomenological parameters in the Navier-Stokes equations can be obtained by a detailed molecular simulation (Brownian dynamics & Langevian dynamics for diffusion coefficients, molecular dynamics for force calculations). The transport of charge, such as ion transport, can be studied by the Poisson-Nernst-Planck (PNP) formulation recently developed. The combination of the phenomena of neutral bio-fluidic transport and charge transport can be simulated by the phenomenological hydrodynamic equations, which provide a more detailed description of transport phenomena than the PNP formulation. In addition to the electrostatics, the energy and momentum exchanges are explicitly included, and the system is no longer assumed to be in the over-damped limit of frequent microscopic collisions like in PNP.

The biological fluidic transport properties of the proposed hybrid devices of ion channels and nanotubes can be simulated. Simulations of individual component of ion channel and nanotube separately have been conducted, therefore, an integration is needed to study the coupled hybrid system. The phenomenological relations of ion channel transport can be applied with those physical principles of nanotubes together to yield phenomenological laws of the proposed hybrid system of ion channels and nanotubes.

5. Constructing a two-dimensional array hybrid detection system

It is possible to construct a two-dimensional array detection system can be constructed by putting carbon nanotubes in a well-defined two dimensional system as a template for docking of biological molecules (NOT BY PUTTING ION CHANNELS IN A TWO-DIMENSIONAL ARRAY, which is hopeless). It is difficult to manipulate ion channels and arrange them to form a well-defined two-dimensional array, but Carbon nanotubes can be formed on a well-defined two dimensional array.

Therefore, the carbon nanotube array will be the template, which an individual ion channel (of same type or of different type for different sensing tasks) can be "welded" one-by-one onto each of the nanotube to form an array of hybrid devices. Note that the ion channels, "welded" onto each of the nanotube, can be of different type with very different selectivity to detect unique bio-signatures for various target molecules of interest. A parallel multi-channel detection system with high-throughput can be built by making many such hybrid devices working simultaneously in a two-dimensional array.

6. Constructing a nano-scale basic current amplification circuit

It is possible to construct a nano-scale ionic current based current amplification circuit acting like a basic semiconductor transistor, based on a biological ion channel and a carbon nanotube. As voltage and macro-molecules can be attached to a carbon nanotube to alter its electronic conducting properties, a Y-shaped carbon nanotube can be grown to make a three-terminal coupled transport system, like a semiconductor transistor. Again, ion-channels can be attached to one or many ends of a Y-shape carbon nanotube to yield the desired coupling properties.

7. Constructing an active transport system

It is possible to use hybrid devices in facilitating biological active transport system to deliver ATP (energy storage source) and making a possible molecular battery. A hybrid device can be built with nanotubes with a natural active transport bio-

Table I. Comparison of Semiconductor Devices with Ion Channels

	Semiconductor Devices	Ion Channels
Dimension or Length (cm)	10^{-2}	10^{-7}
Voltage (Volts)	10	10^{-1}
Electric Field (V/cm)	10^{3}	10^{6}
Velocity (cm/sec)	$\sqrt{(\frac{8k_BT}{\pi m})} \sim 10^{7}$	10^{5}
Carrier Concentration (M/liter)	10^{-5}	10^{-3}
Permanent Charge or Doping Concentration (M/liter)	10^{-2}	1

logical system, which drives transport phenomena by the consumption of chemical energy stored in ATP. A reversal of this cycle by other driving forces (demonstrated recently possible in proton-pump systems, for example), such as mechanical pressure or electrical potential gradients, will make this hybrid device a rechargeable micro-battery, capable of storing different forms of energy.

In summary, we propose to develop ion channel/nanotube hybrid devices designed to act as innovative and revolutionary sensors, devices and miniature systems with the support of computational modeling, experimental and engineering foundations of ionic channels/nanotubes. The prospect of the combination ionic channels with engineered nanotubes to build hybrid devices and systems could lead to radical advances in the integration of sensing, manipulation of biological matters, drug delivery, cell-cell signaling, digital computation and information technology.

4. CONCLUSION

The rich properties of engineered carbon nanotubes will find themselves in wide range of biological and medicinal applications. Combining the natural biological molecular recognition process by natural proteins (ion channels, for example) with the properties of carbon nanotubes, an integrated new class of hybrid devices of carbon nanotubes and biological proteins will have versatile unprecedented selectivity, sensitivity, efficiency, and controllability.

5. ACKNOWLEDGEMENT

DPC is grateful to Profs. Michael A. Stroscio and Mitra Dutta for their encouragement and to Bob Eisenberg for working together years on ion channel permeation problem.

6. REFERENCES

1. Bertil Hille. *Ionic Channels of Excitable Membranes.* Sinauer Associates Inc., Sunderland, Massachusetts, second edition, 1992.

2. Lawrence H. Pinto, L. J. Holsinger, and R. A. Lamb, Influenza virus M_2 protein has ion channel activity, *Cell*, **69**, 517–528 (1992).

3. D. Dawson, editor. *Ion Channels and Genetic Diseases.* Society of General Physiologists. Rockefeller University Press, New York, 1995.

4. S. W. Cowan, T. Schirmer, G. Rummel, M. Steiert, R. Ghosh, R. A. Pauptit, J. N. Jansonius, and J. Rosenbusch, Crystal structures explain two functional properties of two E. coli porins, *Nature*, **358**, 727–733 (1992).

5. Denis Jeanteur, Tilman Schirmer, Didier Fourel, Valerie Simonet, Gabriele Rummel, Christine Widmer, Jurg P. Rosenbusch, Franc Pattus, and Jean-Marie Pagès, Structural and functional alterations of a colicin-resistant mutant of OmpF porin from Escherichia coli, *Proc. Natl. Acad. Sci. USA*, **91**, 10675–10679 (1994).

6. Roger Sayle. *RasMol: Molecular Graphics Visualisation tool.* Biomolecular Structures Group, Glaxo Wellcome Research & Development, Stevenage, Hertfordshire, UK, 1995.

7. R. Misra and S. A. Benson, Genetic identification of the pore domain of the OmpC porin of Escherichia coli K-12, *J. Bacteriol.*, **170**, 3611–3617 (1988).

8. G. Meissner, Ryanodine activation and inhibition of the Ca^{2+} release channel of sarcoplasmic reticulum, *J. Biol. Chem.*, **261**, 6300–6306 (1986).

9. J. Smith, T. Imagawa, J. Ma, M. Fill, K. Campbell, and R. Coronado, Purified ryanodine receptor from rabbit skeletal muscle is the calcium-release channel of sarcoplasmic reticulum, *J. Gen. Physiol.*, **92**, 1–26 (1988).

10. B. Jap and P. Walian, Biophysics of structure and function of porin, *Quart. Rev. Biophys.*, **23**, 367–403 (1990).

11. H. Nikaido, Transport across the bacterial outer membrane, *J. Bioenerg. Biomembr.*, **25**, 581–589 (1993).

12. Jurg Rosenbusch. Porins In *Phosphate in Microorganisms. Celluar and Molecular Biology*, A. Torriani-Gorini, E. Yagil, and S. Silver, editors, pages 329–334. ASM Press, Washington, DC, 1994.

13. Xuanqing Jiang, Marvin A. Payne, Zhenghua Cao, Samuel B. Foster, Jimmy B. Feix, Salete M. C. Newton, and Phillip E. Klebba, Ligand-specific opening of a gated-porin channel in the outer membrane of living bacteria, *Secience*, **276**, 1261–1264 (1997).

14. D. A. Doyle, J. M. Cabral, Richard A. Pfuetzner, A. Kuo, J. M. Gulbis, S. L. Cohen, B. T. Chait, and R. MacKinnon, The structure of the potassium channel: Molecular basis of K^+ conduction and selectivity, *Science*, **280**, 69–77 (1998).

15. Bert Sakmann and Erwin Neher, editors. *Single Channel Recording.* Plenum Press, New York, 2nd edition, 1995.

16. Duan P. Chen, Le Xu, Ashutosh Tripathy, Gerhard Meissner, and Bob Eisenberg, Permeation through the calcium release channel of cardiac muscle, *Biophys. J.*, **73**, 1337–1354 (1997).

17. V. Barcilon, D. P. Chen, R. S. Eisenberg, and M. A. Ratner, Barrier crossing with concentration boundary conditions in biological channels and chemical reactions, *J. Chem. Phys.*, **98**, 1193–1212 (1993).

18. Masakatsu Watanabe, J. Rosenbusch, T. Schirmer, and Martin Karplus, Computer simulations of the OmpF porin from the outer memebrane of Escherichia coli, *Biophys. J.*, **72**, 2094–2102 (1997).

19. S.-W. Chiu, S. Subramaniam, E. Jakobsson, and J. A. McCammon, Water and polypeptide conformations in the gramicidin channel: a molecular dynamics study, *Biophys. J.*, **56**, 253–261 (1989).

20. Johan Åqvist and Arieh Warshel, Energetics of ion permeation through membrane channels, solvation of Na^+ by gramicidin A, *Biophys. J.*, **56**, 171–182 (1989).

21. Benoît Roux and Martin Karplus, Ion transport in a model gramicidin channel: structure and thermodynamics, *Biophys. J.*, **59**, 961–981 (1991).

22. Wonpil Im and Benoit Roux, Ion and counterions in a biological channel: A molecular dynamics simulation of ompf porin from escherichia coli in an explicit membrane with 1m kcl aqueous salt solution, *J. Mol. Biol.*, **319**, 1177–1197 (2002).

23. Wonpil Im and Benoit Roux, Ion permeation and selectivity of ompf porin: A theoretical study based on molecular dynamics, brownian dynamics and continuum electrodiffusion theory, *J. Mol. Biol.*, **322**, 851–869 (2002).

24. Duan P. Chen, Le Xu, Ashutosh Tripathy, Gerhard Meissner, and Bob Eisenberg, Selectivity and permeation through the calcium release channel of cardiac muscle: Monovalent alkaline metal ions, *Biophys. J.*, **76**, 1346–1366 (1999).

25. Duan P. Chen, Le Xu, Bob Eisenberg, and Gerhard Meissner, Calcium ion permeation through the calcium release channel (ryanodine receptor) of cardiac muscle, *J. Phys. Chem.*, **107**, 9139–9145 (2003).

INDEX